CU01642706

N

The game of tennis raises many challenging questions to a statistician. Is it true that serving first in a set gives an advantage? Or serving with new balls? Is the seventh game in a set particularly important? Are top players more 'stable' than other players? Do real champions win the big points? These, and many other questions, are formulated as 'hypotheses' and tested statistically. This book discusses how the outcome of a match can be predicted (even while the match is in progress), which points are important and which are not, how to choose an optimal service strategy, and whether a 'winning mood' actually exists in tennis. Aimed at readers with some knowledge of mathematics and statistics, the book uses tennis (Wimbledon in particular) as a vehicle to illustrate the power and beauty of statistical reasoning.

Franc Klaassen is Professor of International Economics at the University of Amsterdam. After obtaining masters degrees in econometrics and economics and a PhD at Tilburg University, he moved to the University of Amsterdam in 1999. Klaassen is a fellow of the Tinbergen Institute and was a visiting fellow at the University of Wisconsin-Madison. His main research interests are the empirical analysis of international economics and finance, fiscal policy, and sports, mainly tennis, on which he has published widely. He is an enthusiastic tennis player and, as a junior, was selected to train with the Royal Dutch Lawn Tennis Association for nine years.

Jan R. Magnus is Emeritus Professor at Tilburg University and Visiting Professor of Econometrics at the VU University Amsterdam. He studied econometrics and philosophy at the University of Amsterdam, where he obtained his PhD in Economics. He worked at the Universities of Amsterdam, Leiden, and British Columbia before moving to the London School of Economics in 1981. In 1996 he was appointed Research Professor of Econometrics at Tilburg University. Magnus held visiting positions at University of California San Diego, New Economic School of Moscow, European University Institute in Florence, and University of Tokyo, among others. He is author or coauthor of eight books and more than one hundred scientific papers.

Analyzing Wimbledon

The Power of Statistics

Franc Klaassen

Amsterdam School of Economics,
University of Amsterdam, The Netherlands

Jan R. Magnus

Department of Econometrics & Operations Research,
VU University Amsterdam, The Netherlands

OXFORD
UNIVERSITY PRESS

OXFORD
UNIVERSITY PRESS

Oxford University Press is a department of the University of Oxford. It furthers the University's objective of excellence in research, scholarship, and education by publishing worldwide.

Oxford New York
Auckland Cape Town Dar es Salaam Hong Kong Karachi
Kuala Lumpur Madrid Melbourne Mexico City Nairobi
New Delhi Shanghai Taipei Toronto

With offices in
Argentina Austria Brazil Chile Czech Republic France Greece
Guatemala Hungary Italy Japan Poland Portugal Singapore
South Korea Switzerland Thailand Turkey Ukraine Vietnam

Oxford is a registered trade mark of Oxford University Press in the UK and certain other countries.

Published in the United States of America by
Oxford University Press
198 Madison Avenue, New York, NY 10016

© Franc Klaassen and Jan R. Magnus 2014

All rights reserved. No part of this publication may be reproduced, stored in a retrieval system, or transmitted, in any form or by any means, without the prior permission in writing of Oxford University Press, or as expressly permitted by law, by license, or under terms agreed with the appropriate reproduction rights organization. Inquiries concerning reproduction outside the scope of the above should be sent to the Rights Department, Oxford University Press, at the address above.

You must not circulate this work in any other form and you must impose this same condition on any acquirer.

Library of Congress Cataloging-in-Publication Data
Klaassen, Franc.
Analyzing Wimbledon: the power of statistics / Franc Klaassen, Jan R. Magnus.
pages cm
Includes bibliographical references and index.
ISBN 978-0-19-935595-2 (cloth: alk. paper) – ISBN 978-0-19-935596-9 (paperback: alk. paper) 1. Wimbledon Championships (Wimbledon, London, England) 2. Tennis–Statistics. I. Magnus, Jan R. II. Title.
GV999.K58 2014
796.342094212–dc23 2013026355

1 3 5 7 9 8 6 4 2

Typeset by the authors
Printed in the United States of America on acid-free paper

To my parents (FK)
To Eveline de Jong (JM)

Contents

Preface

This is a book about tennis and about statistics. It is possible — as some of our friends predict — that tennis enthusiasts will find the statistics incomprehensible and that statisticians will not be interested in tennis. We hope that our friends are mistaken and that instead the book will encourage people interested in tennis to learn more about tennis and (as a bonus) learn some statistics. We also hope that people with some knowledge of statistics will see how statistical modeling and analysis can be applied to questions in tennis, thus learning more about statistics and (as a bonus) about tennis.

Our own interest in tennis as a field of statistical study started more than fifteen years ago, when watching Wimbledon on television and listening to the commentators. We collected more than twenty statements often made by commentators, such as: it is an advantage to serve first in a set, the seventh game is the most important game in the set, a player is as good as his or her second service, new balls are an advantage to the server, and real champions win the big points. After obtaining the necessary data from IBM, the Official Technology Partner of Wimbledon since 1990, we analyzed these statements and found many to be false. At that point, our analysis was not sophisticated and our interest was only half-serious.

Reflecting more on the possibilities of tennis data, we set ourselves the task of answering four questions: (a) which of the hypotheses on tennis are true and which are not?, (b) do players play every point as it comes, or are they affected by the past (winning mood) and by characteristics of the current point?, (c) how can we forecast who wins a match, not only at the beginning but also during the match?, and (d) is there an optimal service strategy and how close are professional players to this optimal strategy? We answered the first question in Magnus and Klaassen (1999a,b,c; 2008) and the other three questions in Klaassen and Magnus (2001, 2003a, 2009). Our work appeared in newspapers, tennis magazines, and scientific journals, thus catering to a variety of audiences and requiring a presentation of the material at different levels.

For an academic to study tennis or some other sport may require a defense or at least an explanation. We offer no defense of our interest in tennis. We could argue (and we shall expand on this later) that sport statistics as a discipline is useful, because the data are clean and tell us something about human behavior. But the truth of the matter is that we only discovered this later and that we would have studied tennis even if such additional benefits had been absent.

We can, however, explain why we wrote this book. Our first aim was to combine the material that we had discussed in our previous papers with many new results, all within one framework suitable for a specific audience, going slowly from simple to more sophisticated. Our second aim was to show that statistics, when applied carefully, can provide insights that cannot be obtained otherwise; in other words, to demonstrate the power of statistics.

The typical reader we have in mind has some knowledge of and interest in mathematics and statistics, for example at the level of a third- or fourth-year undergraduate in the US college system, and of course some interest in tennis. The book should also be of interest to tennis enthusiasts with less mathematical background. Chapter 13 (the final chapter) is written especially for this audience. It can be read separately from the rest of the book and contains no mathematics.

We hope that our book will prove useful for teaching students the power and scope (and the limitations) of statistics, will provide new insights to tennis enthusiasts and commentators, and will even tell readers something about human behavior more generally. If we inspire the reader to share at least some of our enthusiasm for tennis or statistics or both, then we shall consider our mission accomplished.

Amsterdam, September 2013

Acknowledgements

We are grateful to IBM UK and The All England Lawn Tennis and Croquet Club at Wimbledon for their kindness in providing us with Wimbledon data at point level. Without this initial data set, there would have been no project, no papers, and no book. In addition, we were given summary statistics for all four grand slam tournaments, and we thank IBM UK and the International Tennis Federation for their kind cooperation.

Although most of the material in this book has been rethought and recalculated, we have drawn freely on our earlier publications. We are grateful to the following publishers, associations, and societies for permission to use material published earlier by us in their journals: *Kwantitatieve Methoden*, *Psychologie*, Royal Statistical Society (*The Statistician*), Taylor & Francis Ltd. (*Journal of Applied Statistics*), American Statistical Association (*Journal of the American Statistical Association*), *European Journal of Operational Research*, *Medicine and Science in Tennis*, *AENORM*, *Medium Econometrische Toepassingen (MET)*, *STAtOR*, *Tennis Magazine*, and *Journal of Econometrics*.

Other work appeared earlier in a conference volume of the International Statistical Institute (ISI) Meetings in Istanbul (1997), and as chapters in the following three books: *Tennis Science & Technology* (Eds S.J. Haake and A. Coe), Blackwell Science, Oxford; *Tennis Science & Technology 2* (Ed S. Miller), International Tennis Federation, London; and *Statistical Thinking in Sports* (Eds J. Albert and R.H. Koning), Chapman & Hall/CRC Press, Boca Raton, FL. We thank the publishers for permission to use material from these chapters in the current book.

Much thanks are due to Josette Janssen for editorial assistance, to Jozef Pijnenburg for helping us with the layout and answering all our LaTeX questions, to Eveline de Jong for her invaluable help to transform Chapter 13 into a palatable text for the non-mathematical reader, and to three exceptionally knowledgable and sympathetic anonymous reviewers.

Almost twenty years have passed since we started this project in 1994. During this time we talked to many people, attended workshops, gave presentations, and received anonymous referee reports. Our work has greatly benefitted from the constructive comments and knowledge of colleagues and friends, and we thank in particular: Roel Beetsma, Jan Boone, Maurice Bun, Erwin Charlier, Eric van Damme, Tijmen Daniëls, Dmitry Danilov, George Deltas, Bas Donkers, Martin Dufwenberg, Kees Jan van Garderen, Ronald van Gelder, Noud van Giersbergen, Wouter den Haan, Harry Huizinga, Masako Ikefuji, Henk Jager, Frank de Jong, Philip Jung, Frank Kleibergen, Ruud Koning, Siem Jan Koopman, Knox Lovell, Carl Morris, Theo Offerman, Geoff Pollard, Thijs ten Raa, Ward Romp, Arthur van Soest, Keith Sohl, Joep Sonnemans, Mark Steel, Koen Vermeylen, Tom Wansbeek, Alan Woodland, Arnold Zellner, and Katia Zhuravskaya. We have been fortunate indeed to have such supportive colleagues.

1 Warming up

Suppose you are watching a tennis match between Novak Djokovic and Roger Federer. The commentator says: 'Djokovic serves first in the set, so he has an advantage'. Why would this be the case? Perhaps because he is then 'always' one game ahead, thus creating more pressure on the opponent. But does it actually influence him and, if so, how? Now we come to the seventh game, according to some the most important game in the set. But is it? Federer serves an ace at breakpoint down (30-40). Of course! Real champions win the big points. But they win most points on service anyway, including the unimportant points. Do the real champions outperform on big points or do weaker players underperform, so that it only *seems* that the champions outperform? (The latter will turn out to be the case.) Then Djokovic serves with new balls, assumed to be an advantage. But is it really? Next, Federer wins three consecutive games. He is in a winning mood, the momentum is on his side. But does a 'winning mood' actually exist in tennis? All these questions, and many more, will be discussed in this book.

Wimbledon

Vulcanized rubber balls that bounce well on grass were not available until around 1870, and they were a necessary ingredient for the invention of lawn tennis (played outside), which was based on the then existing game of 'real tennis' (played inside). The invention is usually attributed to Major Walter C. Wingfield, who patented his new recreation in 1874 and called it 'sphairistike', an ancient Greek term meaning 'skill of playing with a ball'. The name was never popular, not least because only those few people well versed in ancient Greek knew how to say it. Luckily, Major Wingfield also called his new recreation 'lawn tennis', and this immediately caught on.

The All England Croquet Club at Wimbledon was founded in the Summer of 1868. Lawn tennis was first played at the club in 1875, when one lawn was set aside for this purpose. In 1877 the club was retitled The All England Croquet and Lawn Tennis Club. In 1882, croquet was dropped from the name, as tennis had become the main activity of the club, but in 1889 it was restored to the club's name for sentimental reasons, and the club's name became The All England Lawn Tennis and Croquet Club.

The first tennis championship was held in July 1877 (men's singles only) with twenty-two players. Spencer Gore became the first champion and won the Silver Challenge Cup and twelve guineas, no small sum (about £800 or $1240 in today's value), but rather less than the £1,150,000 ($1,785,000) that each of the 2012 champions Roger Federer and Serena Williams received.

In 1884 the women's singles event was held for the first time. Thirteen players entered this competition and Maud Watson became the first women's champion, receiving twenty guineas and a silver flower basket. (William Renshaw, the 1884 men's singles champion, received thirty guineas.)

For more than a century The Championships at Wimbledon have been the most important event on the tennis calendar. Currently 128 men and 128 women participate in the main draw of the men's and women's singles, competing over seven rounds. Especially because of television broadcasts, tennis is no longer only a recreation, but has become a sport viewed by millions of people all over the world. And everybody has ideas about tennis: players, viewers, journalists, and television commentators.

Commentators

Tennis is one of the most difficult sports to commentate on. In football the commentator can provide the name of the player in ball possession. In snooker the commentator can suggest possible solutions to the problem on the table. In running and swimming there is a continuum of time in which the event takes place, maybe short (ten seconds), maybe long (two hours), but uninterrupted.

In a tennis match, all these advantages are absent. To provide the name of the player hitting the ball is ridiculous, to mention the score is often redundant, to make technical comments can occasionally be illuminating, but the most serious problem is that most of the time *nothing happens.*

In a men's singles tennis match at Wimbledon, one point lasts about five seconds. One game takes about six points, or thirty seconds. One set takes about ten games, or five minutes. And the match may take four sets, or twenty minutes. In reality, the match does not take twenty minutes but perhaps three hours. Only 10% of viewing time is taken up by actual play; the rest of the time must be filled by the commentator.

This is no easy job. But the job is lightened when the commentator can rely on a number of 'idées reçues', commonly accepted ideas: for example that serving with new balls provides an advantage, that a 'winning mood' exists, or that top players must 'grow into the tournament' and that they perform particularly well at the 'big' points. Some of these ideas are true but many of them are false, and one of the purposes of this book is to decide which of these idées reçues are true and which are not.

An example

Let us consider the hypothesis that the player who serves first in a set has an advantage, a typical example of an idée reçue, based on the idea that the player who serves first experiences less pressure. We have data on more than one thousand sets played at Wimbledon and we can simply calculate how often the player who served first also won the set. This statistic shows that there is a slight advantage in the first set, but no advantage in the other sets. On the contrary, in the other sets there is a disadvantage: the player who serves first in the set is more likely to lose the set than to win it. This is surprising. What could be the explanation? Perhaps it is different for the women? But no, the same pattern occurs in the women's singles.

The explanation is that the player who serves first in a set (if it is not the first set) is usually the weaker player. This is so, because (a) the stronger player is more likely to win the previous set, and

(b) the previous set is more likely won by serving the set out than by breaking serve. Therefore, the stronger player typically wins the previous set on service, so that the weaker player serves first in the next set. The weaker player is more likely to lose the current set as well, not because of a service (dis)advantage, but because he or she is the weaker player.

This example shows that we must be careful when we try to draw conclusions based on simple statistics. In this case, the fact that the player who serves first in the second and subsequent sets often loses the set is true, but this concerns weaker players while the hypothesis concerns all players. If we wish to answer the question of whether serving first *causes* a (dis)advantage, we have to control for quality differences. If we do this correctly, then we find that there is no advantage or disadvantage for the player who serves first in a set; in other words, it does not matter who serves first in the second or subsequent sets. But in the first set it *does* matter (we'll see later why), so it is wise to elect to serve after winning the toss.

But how should we account for differences in quality? The players' positions on the world-ranking lists obviously give an indication. But how good is this indication? And there are also other aspects of quality — such as 'form of the day' — which are not captured by the ranking and cannot even be observed. How do we account for these? All these issues will be dealt with in this book.

Correlation and causality

In dealing with these and other questions we need to know a little about statistics. Statistics does not have a good reputation. Many people agree with Mark Twain's phrase 'lies, damned lies, and statistics'. One well-known cause of ridicule is the confusion between correlation and causality. There is a high correlation between eating much garlic (as in Greece, Italy, and Portugal) and having a high government budget deficit. Maybe eating garlic causes the deficit? A new law forbidding garlic consumption would then solve the current economic crisis. There is also a high correlation between reading skills of children and their shoe size, but it would be hazardous to conclude that children with big feet are more intelligent.

There are many types of fallacies associated with correlation:

for example reverse causation (the more firemen are fighting a fire, the bigger the fire is likely to be — hence, more firemen cause bigger fires) and common cause (reading skills and shoe size for young children have a common cause, namely the age of the child).

Sometimes correlation is a coincidence. If we consider many data series, then some will be correlated without any common cause. Apparently, near-perfect correlation exists between the death rate in Hyderabad, India, from 1911 to 1919, and variations in the membership of the International Association of Machinists during the same period. But this does not imply a causal relationship.

All these examples are based on statistics. Using data on garlic consumption and budget deficits, we do find a positive correlation. This is not a mistake; it is a true reflection of what the data tell us. The mistake lies in the confusion between correlation and causality, and, more generally, in a poor understanding of how statistics can help us to understand the world. We should not blame statistics when someone suggests solving the economic crisis by forbidding garlic consumption. We should blame the ignorance of the person who suggests this.

Why statistics?

In this book we hope to demonstrate that, while bad use of statistics can provide misleading and incorrect answers, good use of statistics can help to clarify phenomena and thus provides important tools for making decisions in uncertain situations. The previous examples, including the possible (dis)advantage of serving first in a set, show that using correct statistics in an incorrect way produces incorrect conclusions. This book will provide more examples of this incorrect use of statistics in tennis, but its main focus will be to show how to use statistics correctly, leading to credible conclusions.

The word 'statistics' has two meanings. One meaning is a collection of data characteristics, usually averages, as produced by a data-collecting agency such as a national statistical office. These statistics would include information about how tall people are, life expectancy, fertility rates, and so on, and they are called *descriptive* statistics.

A second meaning is the mathematical science pertaining to the collection, analysis, interpretation, and presentation of data.

It is this meaning that we are primarily interested in, because it provides the tools needed to analyze data — for example the tools to distinguish causality from correlation. This second type is called *mathematical* statistics.

Statistics on tennis typically include such items as how often a player who serves first in a set also wins the set, the number of double faults, and the percentage of first serves in. These are descriptive statistics, and they are useful ingredients. Our interest, however, is not in the descriptive statistics themselves, but in what we can learn from them. This learning process involves different types of analysis based on mathematical statistics, and we shall develop and use more and more sophisticated methods as we go along.

Mathematical statistics also provides insights that go beyond what we infer directly from the data. When only few data are available we need to make more assumptions; that is, we need a stronger model. Consider, for example, the epic match at Wimbledon 2010, where John Isner defeated Nicolas Mahut 70-68 in the final set. This match is unique — there is no match like it in any tennis data set. Still, we can build a realistic model and calculate precisely how exceptional the match was. A second example is forecasting the winner of a match while the match is in play. Under appropriate and realistic assumptions, credible forecasts can be obtained.

Sports data and human behavior

Studying sports is not only of interest to those interested in sports. There is a second (some would say a first) interest, namely the study of human behavior. In (professional) sports the players' objectives are clear: they want to win. The incentives to win are strong and the players are highly trained. In everyday life, people differ much more. Some pupils are eager to score high grades at school, while others just want to pass with minimal effort. Employees in a firm have different tasks and they differ in terms of experience. Many of these differences are difficult to observe, thus hampering accurate inference in psychology, economics, and related disciplines. In sports analysis there is less unobserved heterogeneity, thus allowing more accurate inference.

Moreover, sports data are clean — there are few errors in the data — and the data collection is transparent and can be checked. Data in economics, psychology, and many other sciences are often messy and ambiguous. To work with clean data provides a great opportunity for scientists in these fields, and a welcome change from normal practice. If results do not come out the way they 'should', then this cannot be blamed on the (clean) data. There must be a 'real' explanation and we have to find it. Maybe our preconceived idea is wrong or maybe we have not applied the correct statistical method.

Given the abundance of clean sports data we can try and study human behavior in an *indirect* way. Let us give three examples. Suppose we wish to study whether judges and juries are influenced by social pressure. Useful data from the law courts are not available, so we cannot directly study the possibility of favoritism in the courts of law under social pressure. But we can study favoritism indirectly by considering football (soccer) matches and asking whether referees favor the home team, for example by shortening matches in which the home team is one goal ahead and lengthening matches in which the home team is one goal behind at the end of regular time. It turns out that referees tend to do this, thus favoring the home team.

A second example is the question of whether people become more cautious when pressure mounts. This too can be analyzed indirectly using sports data. In tennis, some points are more important than others. Do players behave differently at the key points? They do: they play safer at important points. This teaches us something about human behavior, and may have implications outside tennis, for example in economics. If salaries of agents working in the financial sector contain not only a bonus but also a substantial malus component, then the consequences of their activities matter in both directions (like winning *or* losing a tennis match). The behavior of professional tennis players suggests that financial agents will then pursue safer actions, reducing the possibility of a banking crisis.

Finally, as a third example, we can ask whether people become less cautious when they are in a winning mood? In tennis language: does a successful spell result in taking more risks, for example a riskier service? It does.

Why tennis?

If we wish to use sports data to analyze human behavior, then tennis is a happy sport to choose. There are several reasons for this. A tennis researcher has to model only one player (in singles matches, which we study). There are no complications caused by intra-team interactions and player substitutions, which could affect the quality of a team and the style of play, as happens in basketball, hockey, volleyball, and other team sports. But there is interaction with the opponent, so strategic behavior can be studied, for example a player's decision to mix the direction and speed of service depending on the performance of the receiver.

The quality of a tennis player is measured by the world ranking, and this provides a good indication of quality. It allows the researcher to control for an important aspect of differences between players, so that inference regarding the question of interest becomes more accurate, as we shall see throughout this book. A proper quality measure makes it also possible to study other questions, for example whether service efficiency and mental stability are related to the quality of a player.

Each tennis match generates a lot of data: many points, many games, many services, and so on. Some sports, like basketball, share this feature, but many sports don't. In football (soccer), for example, there are few goals and few corner kicks. In addition, tennis allows the server to serve twice (first and second service), an exceptional situation compared to other sports, doubling the amount of information on a player's service strategy. Moreover, men's and women's tennis are relatively similar. In fact, men and women play the four grand slam tournaments (Australian Open, Roland Garros, Wimbledon, and the US Open) together. This generates not only more observations and more possibilities to check the robustness of conclusions, but it also makes it possible to study gender differences. All these features will be exploited in this book.

Scoring in tennis is almost objective, much more so than in football for example. It has recently become even more objective by the introduction of the Hawk-Eye technology, allowing the player to make three unsuccessful 'challenges' of the umpire's decision per set in any match using Hawk-Eye, plus one more in the tiebreak. Tennis is one of only a few sports with such a system. It enables a researcher to study the quality of strategic decisions by humans, here by testing whether tennis players make optimal challenge decisions.

Tennis has a tri-nested scoring system. Although many sports consist of successive points, in tennis the points become games, the games become sets, and the sets become the match. This allows for a separation into three levels, facilitating the statistical analysis. It also creates additional quantifiable differences between points, as some points are more important than others, which helps to analyze questions such as how humans behave under stress. In summary: tennis is ideal.

Structure of the book

In this book we examine various types of tennis-related questions: from hypothesis testing to forecasting the winner during a match, to strategic service decisions. We always present the required statistical methods from simple to advanced. Sometimes simple methods suffice, sometimes not. Sometimes conclusions change with increased complexity, sometimes they do not; when conclusions change, we shall comment on the reasons for the change.

The book contains thirteen chapters of which Chapter 1 is the introduction and Chapter 13 provides a non-mathematical summary of all hypotheses discussed in the book. The remaining eleven chapters contain the body of the book and can be summarized as follows.

- Chapters 2–4 discuss our computer program and how to use it in forecasting matches, and introduce the concept of 'importance'. No data are used in these three chapters.

- Chapters 5–9 use point-by-point data from about five hundred Wimbledon matches to study tennis at match level.
    - Chapter 5 introduces the data;
    - Chapters 6–8 develop the statistical method and apply this method to study a player's 'quality' and first and second service, and to test a number of hypotheses;
    - Chapter 9 builds on Chapter 8 and studies the strategy and efficiency of the service.

- Chapters 10–12 deepen the analysis by studying points within a match and introducing dynamics. This allows us to study big points, winning mood, and other aspects of dynamics.

Three appendices accompany this book. In Appendix A we briefly summarize tennis rules and tennis terms; in Appendix B we provide a list of symbols used in this book; and in Appendix C we describe the data and the software, and direct the reader to our websites for further information, including derivations of some of the mathematical formulas.

Further reading

Tennis is played by more than seventy-five million people worldwide and is one of the major global sports (Pluim *et al.*, 2007). The International Tennis Federation (ITF), founded in 1913, is the governing body of tennis and determines the 'Rules of Tennis'. The latest Rules can be downloaded from the ITF website. Two other important organizations are the Association of Tennis Professionals (ATP) for the men and the Women's Tennis Association (WTA) for the women. Both organizations maintain informative websites containing information on players, rankings, tournaments, past results, and so on. For historical tennis information see Collins (2010); for Wimbledon facts see Little (1995).

Tennis has been studied from many different angles, and most of these have been summarized in review articles providing many additional references. Brody *et al.* (2002) cover the physics of rackets, strokes, strings, tennis balls, and courts. The tennis ball, in particular, has received interest given its aerodynamics, and has even been the subject of wind tunnel experiments at NASA. Isaac Newton studied a predecessor of the tennis ball in 1672; see Mehta *et al.* (2008). The 'tennis elbow' is the best-known tennis injury, but there is also the 'tennis leg', and much more. Pluim and Safran (2004) provide sports medicine guidance on how to play healthy tennis, and Miller (2006) considers tennis equipment and its relationship to common tennis injuries. Elliott *et al.* (2009) outline the mechanical basis of stroke development, useful for training and coaching, while Crespo *et al.* (2006) summarize the role of psychology. Pollard and Meyer (2010) describe operations research methods to improve the scoring system. The early statistical literature on tennis is well summarized in Croucher (1998).

Most studies focus on singles matches, as we do, but there exists a doubles literature as well. Anderson (1982) finds that the female player in mixed doubles accounts for a larger part of the team's success than the man, while Clarke (2011) shows how clubs can rate their (non-elite) players in doubles competitions.

Lake (2011) describes behavioral etiquette in tennis from 1870 to 1939, and how this has affected the choice of shot. Volleying in the 1870s, for example, was considered ungentlemanly. Even 'grunting' (making noises while serving or hitting the ball) has been studied, by Sinnett and Kingstone (2010).

Although there is currently no book on the statistical analysis of tennis, there are studies covering other sports or sports in general. Humphreys (2011) presents a method for analyzing baseball fielding statistics. This enables him to put players from different eras on equal footing, so that he can rank the best fielders at each position throughout baseball history. An anthology of statistics in sports is given in Albert *et al.* (2005). Shmanske and Kahane (2012) discuss many articles on the interaction between economics and sports. Economic analysis helps us to understand sports institutions, and quality data on sports help economists to study topics such as discrimination, salary dispersion, and antitrust policy. British football is studied in Dobson and Goddard (2011). The social pressure example discussed on page 7 is taken from Garicano *et al.* (2005), who studied two seasons of the Spanish football competition.

2 Richard

In this chapter we introduce a computer program, called *Richard*, named after Richard Krajicek, the only Dutchman ever to win the Wimbledon singles title, in 1996. *Richard* calculates tennis probabilities, in particular the probability of winning a point, game, set, or match. We discuss the application of *Richard* to the occurrence of upsets (do these happen more often in the men's singles than in the women's singles?) and long matches (the epic Wimbledon 2010 match between John Isner and Nicolas Mahut). We also analyze rule changes.

Meeting *Richard*

It will be convenient to give a name to the two players in a singles tennis match, and we shall call them $\mathcal{I}$ and $\mathcal{J}$, the calligraphic variants of the letters i and j. For a tennis match between the two players $\mathcal{I}$ and $\mathcal{J}$, *Richard* calculates the probability that $\mathcal{I}$ wins the match. Of course, *Richard* also calculates the probability that $\mathcal{J}$ wins the match, because if $\mathcal{I}$ has, say, a 70% chance of winning, then $\mathcal{J}$'s chance is 30%. In reality we do not observe this probability. The probability which *Richard* computes is an approximation of the true unobserved probability, based on certain simplifying assumptions. Whether these assumptions are justified needs to be (and will be) checked continuously.

To compute the match-winning probability, *Richard* requires two key inputs: the point-winning probabilities p_i and p_j. More precisely, p_i denotes the probability that $\mathcal{I}$ wins a point on service (against $\mathcal{J}$), and p_j denotes the probability that $\mathcal{J}$ wins a point on service (against $\mathcal{I}$). The main assumption underlying *Richard* is that these two probabilities do not change during one match. A statistician would say that the points served by $\mathcal{I}$ are independent and identically distributed (iid), and so are the points served by $\mathcal{J}$. We formulate this important assumption as a hypothesis.

Hypothesis 1: *Winning a point on service is an iid process.*

Many hypotheses will be considered in this book, and this is the first. Is it a reasonable assumption to make? In amateur tennis, players are often much affected by what happened at the previous point. For example, missing a smash that you should not have missed could very well make you lose the next point as well. The points are then dependent. The more professional the players are, the less dependence one would expect. *Richard* assumes no dependence at all.

Players could also be affected by the score. Maybe they play differently at 40-0 than at 30-40 (breakpoint). When this happens the points are not identically distributed. We assume that points *are* identically distributed. Of course, points have different characteristics (such as the score), but the *probability* of winning a point is the same, at least if hypothesis 1 holds.

The iid assumption is, in fact, more than an assumption; it is also a strategy. The strategy involves not living in the past or in the future, but only in the present. What happened at the previous point is no longer relevant. The score is not relevant. Whom you are playing in the next round is not relevant. Only the current point matters. This is easier said than done. 'One point at a time' may be a cliché, but we shall see later in Chapter 10 that the better a player is, the closer he or she comes to satisfying the iid assumption, that is, to playing one point at a time.

The iid assumption may appear unrealistic, and of course the assumption is not perfectly true. In statistical modeling, however, whether a simplifying assumption is true or not is not the issue. Of course the assumption is not true! What matters is whether the simplified model brings us sufficiently close to what we want to achieve. Albert Einstein expressed this by saying: 'As simple as possible, but not simpler'.

Let us assume then, for the time being, that hypothesis 1 is satisfied. The probability that a player wins the match then depends on the point probabilities p_i and p_j, the type of tournament (best-of-three sets or best-of-five sets, tiebreak in final set or not), the current score, and the current server, but on nothing else. *Richard* calculates the probability of winning the current game (or tiebreak), the current set, and the match. These probabilities are calculated exactly, based on recursive formulas. The program, first published by us in 1995, is flexible and very fast. It is flexible, not only because it allows the user to specify the score and to adjust to the particularities of the tournament, but also because it allows for rule changes. For example, we can analyze what would happen if the traditional scoring rule at deuce is replaced by the alternative of playing *one* deciding point at deuce, or what would happen if the tournament requires four games rather than six to be won in order to win a set (not currently allowed by the official rules). The program is freely available — see Appendix C for details.

From point to game

The simplest example of *Richard* is provided by considering one game in a match between $\mathcal{I}$ and $\mathcal{J}$. A game consists of a sequence of at least four points with the same player serving, and it is won by the first player to have won at least four points with a difference of at least two points. Let us assume that $\mathcal{I}$ is serving and that the probability that $\mathcal{I}$ wins a point is p_i. What is the probability that he or she wins the game? It is

$$g_i = \frac{p_i^4(-8p_i^3 + 28p_i^2 - 34p_i + 15)}{p_i^2 + (1 - p_i)^2}.$$

The derivation of the formula is not completely trivial because of the deuce rule and the uncertainty about how many points there will be in the game. (A full derivation is available on the website accompanying this book; see Appendix C for details.) But it is easy to verify that $p_i = 0$ implies $g_i = 0$, that $p_i = 0.5$ implies $g_i = 0.5$, and that $p_i = 1$ implies $g_i = 1$. In other words, if the server never wins a point on service, he or she never wins a service game; if the server and receiver have equal probabilities of winning a point, they have equal probabilities of winning the game; and if the server wins all service points, he or she wins all service games. More interestingly, we can now calculate that $p_i = 0.6$ implies $g_i = 0.74$,

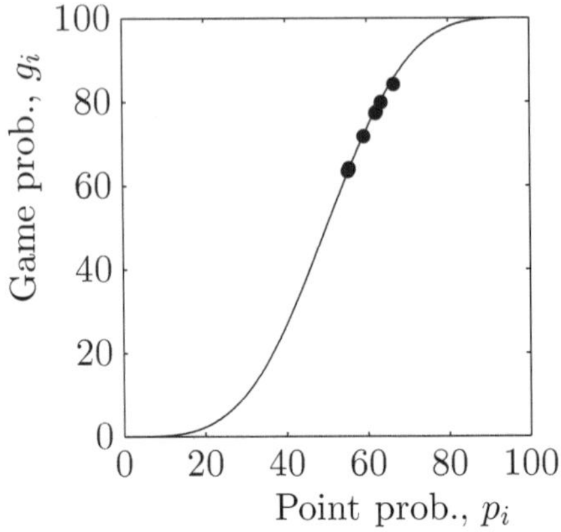

Figure 2.1: *Probability g_i (in %) of winning a service game as a function of the probability p_i (in %) of winning a service point, with 2010 grand slam realizations*

and that $p_i = 0.7$ implies $g_i = 0.90$. Figure 2.1 illustrates the formula by plotting g_i for any p_i.

The fact that $g_i > p_i$ when $p_i > 0.5$ is an example of the *magnification effect*: an advantage at point level leads to a bigger advantage (is magnified) at game level. The S-shaped curve in Figure 2.1 illustrates that the magnification effect is large when p_i is close to one half and small when p_i is close to one. For example, when p_i increases from 0.50 to 0.51, g_i increases by 2.5%-points (from 0.500 to 0.525); when p_i increases from 0.60 to 0.61, g_i increases by 2.0%-points (from 0.736 to 0.756); and when p_i increases from 0.70 to 0.71, g_i increases by only 1.1%-points (from 0.901 to 0.912). When p_i is close to one, then g_i hardly increases any more, because player $\mathcal{I}$ will win the game anyway.

The dots in Figure 2.1 provide some realistic (p_i, g_i) combinations. They represent the relative frequencies of winning a service point and game at the four grand slam tournaments in 2010, as reported in Table 2.1. The eight dots visualize the eight pairs from the table. (Only five are in fact visible because some of the percentages almost coincide.) The relative frequencies can be considered as realizations of the corresponding probabilities p_i and g_i. The practice agrees remarkably well with the theory, as represented by the S-shaped curve, indicating that the iid assumption may not be

	Men		Women	
	Point	Game	Point	Game
Australian Open	62.2	77.2	55.4	63.4
Roland Garros	62.4	77.5	55.7	64.1
Wimbledon	66.7	84.1	59.3	71.7
US Open	63.6	79.8	55.7	63.9

Table 2.1: *Percentage of winning a service point versus winning a service game (in %), 2010*

so bad after all.

We see from Table 2.1 that the service dominance is lowest at the Australian Open and highest at Wimbledon, the only grand slam played on grass. Service dominance is larger for the men than for the women: at Wimbledon 2010, 84.1% of the games was won by the server in the men's tournament compared to 71.7% for the women.

The tiebreak

The tiebreak, invented by James Van Alen in 1965, was introduced at Wimbledon in 1971 following the 1969 first-round match between Pancho Gonzales and Charlie Pasarell, which lasted five hours and twelve minutes and took two days to complete. At the time there were no chairs on court enabling the players to rest when changing ends; these were only introduced six years later, in 1975. Gonzales, then forty-one years old, survived seven matchpoints and won 22-24, 1-6, 16-14, 6-3, 11-9. The tiebreak was introduced to avoid such long matches.

Originally the tiebreak would come in effect when the score in any set, except the final set, reached 8-8 in games. In 1979 Wimbledon changed its rules so that a tiebreak would be played when the score reached 6-6 in games. Tiebreaks are now used in all four grand slam events except in the final set. Only at the US Open is a tiebreak also played in the final set.

The tiebreak is a special type of game, but while in a game four points are needed to win, a tiebreak requires seven. The player whose turn it was to serve in the set serves the first point of the tiebreak. The opponent then serves the next two points and after that the service rotates after every two points. If the score reaches six-points-all, the winner is the first player achieving a two-point advantage.

The tiebreak can also be compared to a set because the service rotates: in a set after every game, in a tiebreak after the first point and then after every *two* points. In a set there could be an advantage in serving first, because the player who serves first in the set is always the first to serve in a new pair of two games in that set; he or she is 'always' one game ahead. But even if there were a serving-first advantage in a set, there can be no such advantage in a tiebreak, because players alternate in being first.

Serving first in a set

Is there an advantage of serving first in a set? Many viewers and commentators believe there is, presumably because the player who serves first is 'always' one game ahead (if he or she keeps winning his or her service games), which would create less pressure on the player serving first or more pressure on the player receiving first, or both. We formalize this statement in the following hypothesis.

Hypothesis 2: *It is an advantage to serve first in a set.*

Readers familiar with hypothesis testing will frown at our formulation. In statistics, we formulate a hypothesis always as the thing we want to reject, not as the thing we hope to accept. So in this case the hypothesis would read: 'It is no advantage to serve first in a set'. The formulation of hypotheses in this book sometimes deviates from statistical rigor in order to gain statements that are more appealing to tennis fans.

In a set, player $\mathcal{I}$ serves in the first game, $\mathcal{J}$ serves in the second game, and so on, until one of them has won six games. If the score reaches 5-5, then two more games are played until the score is either 7-5 (or 5-7) or 6-6. At 6-6 a tiebreak is played, unless it is the final set in which case a two-game difference is required at all grand slam tournaments except the US Open.

To model a game we need only one probability p_i, but to model a set we need both probabilities p_i and p_j. We now employ a trick that is often used in modeling. We create two new parameters

$p_i - p_j$ and $p_i + p_j$. The idea behind creating the new parameters is twofold. First, we realize that p_i and p_j are related to each other: if $\mathcal{I}$ is a top player while $\mathcal{J}$ is not, then we would expect p_i to be large and p_j to be small. In contrast, $p_i - p_j$ and $p_i + p_j$ may be much less related. Second, it may be the case that the probability s_i that $\mathcal{I}$ wins the set (which is what interests us here) depends primarily on one of these two new parameters and very little on the other. We shall see that both ideas hold here.

The interpretation of the new parameters is as follows. The probability p_i that $\mathcal{I}$ wins the point on service depends not only on how well $\mathcal{I}$ serves (and plays in the rest of the rally, $serv_i$), but also on how well $\mathcal{J}$ returns (rec_j). Thus we can write

$$p_i = serv_i - rec_j, \qquad p_j = serv_j - rec_i,$$

and this implies that

$$\begin{cases} p_i - p_j = (serv_i + rec_i) - (serv_j + rec_j)\,, \\ p_i + p_j = (serv_i - rec_i) + (serv_j - rec_j)\,. \end{cases}$$

Hence, $p_i - p_j$ represents the quality difference between the two players, taking both serving and receiving into account. The interpretation of $p_i + p_j$ is less straightforward. It represents the 'serve-receive differential' for both players together. For example, when both players serve well but receive poorly then $p_i + p_j$ will be high.

If $\mathcal{I}$ plays against a weaker player $\mathcal{J}$, then both $serv_i$ and rec_i will be high and both $serv_j$ and rec_j will be low, so that $p_i - p_j > 0$. But there is no reason why $serv_i - rec_i$ should be either large or small. Similarly, we cannot predict the level of $serv_j - rec_j$ and thus of $p_i + p_j$. This suggests that $p_i - p_j$ and $p_i + p_j$ will not be much related; they capture different aspects of the players' qualities. We shall see that $p_i - p_j$ is much more important in our analysis than $p_i + p_j$.

In Figure 2.2 we plot the set-winning probability s_i at the beginning of the set as a function of $p_i - p_j$ for different values of $p_i + p_j$, ranging from 0.8 to 1.6, the empirically relevant interval. What do we see? For given $p_i + p_j$, the probability s_i is a monotonically increasing function of $p_i - p_j$, and this functional relationship is shaped like an S. The collection of all curves for $0.8 < p_i + p_j < 1.6$

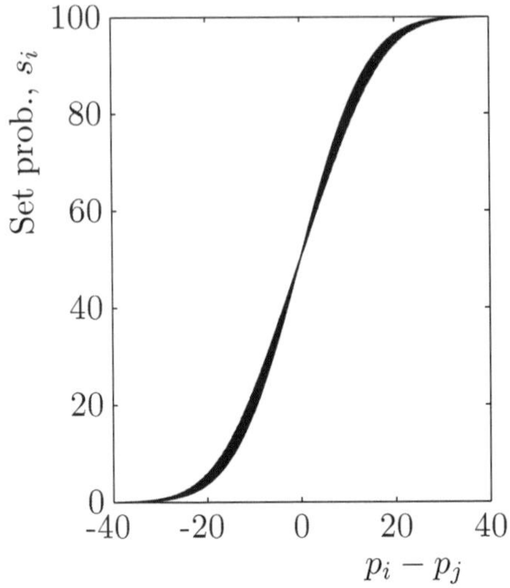

Figure 2.2: *Probability of winning a set at the beginning of the set*

gives the fuzzy S-shaped curve of the figure. We thus see that s_i depends almost entirely on $p_i - p_j$ and only very little on $p_i + p_j$.

In drawing Figure 2.2 we have assumed that $\mathcal{I}$ serves first in the set. Would the graph look different if we had assumed that $\mathcal{J}$ had started? No, it would look precisely the same. If the iid assumption is correct, then there is no advantage in serving first in a set, and hence hypothesis 2 must be false.

But is it? Our conclusion is driven by the iid assumption. We have not yet looked at the empirical evidence, that is, we have not yet employed data. If it is the case that the iid assumption is not correct and that small deviations from iid have a large impact on our question, then the hypothesis could still be true. We shall return to this question in Chapter 11 when we have the relevant data.

During the set

We have just seen that s_i depends almost exclusively on the difference $p_i - p_j$ and hardly at all on the sum $p_i + p_j$. This is at the beginning of a set. Is it also true during a set? At equal scores, such as 4-4, the answer is yes. At unequal scores, however, the dependence on $p_i + p_j$ is stronger, and the two most extreme cases occur at 5-4 and 4-5. Figure 2.3 presents the same relationship as Figure 2.2, but now at 5-4 and 4-5 in games, rather than at 0-0. Player $\mathcal{I}$ served first in the set, so that after nine games $\mathcal{J}$ is serving to stay in the set (when the score is 5-4) or to win the set (when the score is 4-5).

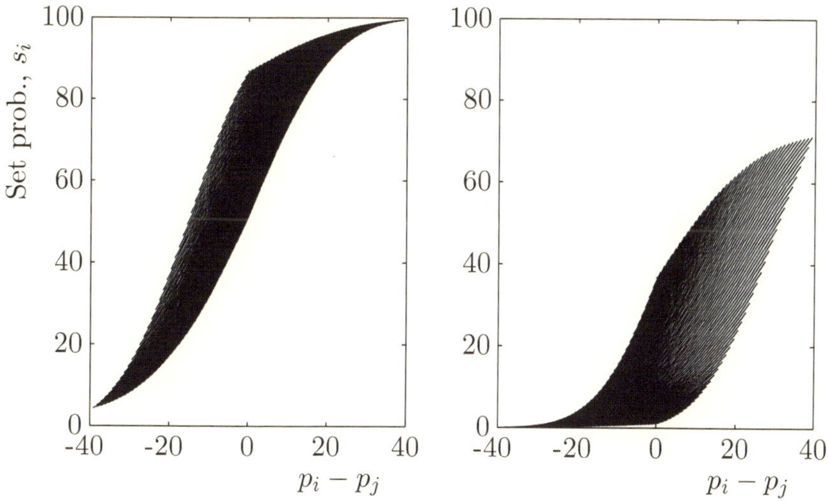

Figure 2.3: *Probability of winning a set at 5-4 (left) and 4-5 (right)*

The dependence on $p_i + p_j$ is now large. Hence, when the score is not equal, a good estimate of $p_i + p_j$ is required to calculate s_i accurately, especially towards the end of a set. Note that the probability in the right panel does not end at $s_i = 1$, because at 4-5 down a certain win by player $\mathcal{I}$ would require that $\mathcal{J}$ has zero probability of winning a point on service ($p_j = 0$). With $p_j = 0$ and $p_i - p_j = 0.4$ we would have $p_i = 0.4$, and such a combination of p_i and p_j is highly unrealistic. A similar reasoning explains why the figure in the left panel does not begin at zero.

Best-of-three versus best-of-five

The usual format for a tennis match is best-of-three: the player who first wins two sets has won. In the majors the women play best-of-three, but the men play best-of-five. In Figure 2.4 we plot the probability m_i that $\mathcal{I}$ wins the match as a function of $p_i - p_j$, for a cluster of values of $p_i + p_j$, calculated at the beginning of the match. Again we see that the dependence on $p_i - p_j$ is much stronger than the dependence on $p_i + p_j$.

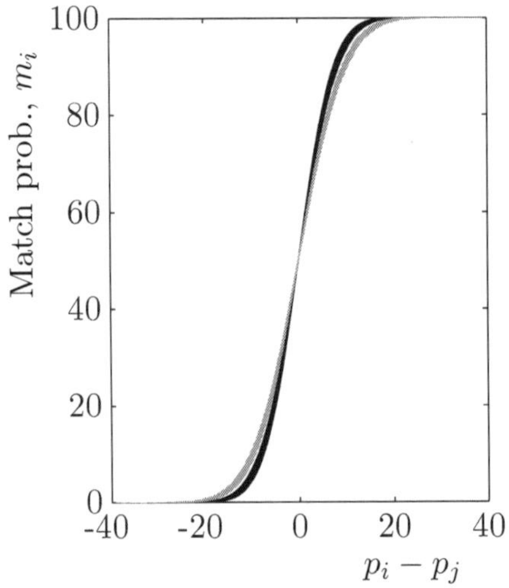

Figure 2.4: *Probability of winning the match, best-of-five (dark) versus best-of-three (light)*

The best-of-five curve (for the men) is always above the best-of-three curve (for the women) when $p_i > p_j$ (and below when $p_i < p_j$). This means that if $\mathcal{I}$ is a better player than $\mathcal{J}$, then, for a given quality difference $p_i - p_j > 0$, the better player has a higher probability of winning in a five-set match than in a three-set match. This makes sense, because the more points have to be played, the higher is the probability that quality will show in the end. It is another example of the magnification effect, introduced on page 16. For example, suppose that $\mathcal{I}$ and $\mathcal{J}$ are of equal strength ($p_i = p_j$) and that $\mathcal{I}$ can raise his or her game so that $p_i - p_j$ is now 0.01. Then the match-winning probability m_i will increase from 0.50 to about 0.55 in a best-of-three match and to about 0.56 in a best-of-five match. We would therefore expect fewer 'upsets' in the men's singles than in the women's singles at grand slam events. But is this what happens?

Upsets

To answer this question we need to define what we mean by an 'upset'. At tennis tournaments a number of players are 'seeded' to avoid having to play against each other in the early rounds. A total of 128 players feature in each grand slam singles event. For many years only the top-sixteen players were seeded. These seedings were sometimes subjective, because seedings were given to players who performed well on the particular surface of that tournament. As a result, some players in the top sixteen would not get a seeding or would be seeded lower than their world ranking, while others who were not in the top sixteen would be seeded. Naturally this annoyed some of the top players. The 2001 Wimbledon championships paved the way for a new and more popular style of seeding, where the top-ranked thirty-two players are seeded for each grand slam tournament, irrespective of their history on a particular surface. This automatic seeding is now standard at all grand slam tournaments except Wimbledon, which still reserves the right to deviate from the official rankings. For the men's singles, this deviation is determined by a formula: ATP points + grass court points during the last twelve months + 75% of points earned from the best grass court tournament during the twelve months before that. But for the women's singles there is no formula, just a recommendation to follow the WTA rankings 'except where, in the opinion of the committee, a change is necessary to produce a balanced draw'.

Intuitively, an upset occurs when an unseeded player beats a seeded player. This is one possible definition of an upset, if we take into account that sixteen players were seeded before 2001, but thirty-two from 2001 onwards. It is easier, however, and equally intuitive to say that an upset has occurred when a top-sixteen player does not reach the last sixteen, and this will be our definition of an 'upset'.

Of the four grand slams, most upsets occur at Wimbledon, both for the men and for the women. Indeed, in 2002 Lleyton Hewitt (seeded 1 and the champion that year) and Tim Henman (seeded 4) were the only top-sixteen seeds to reach the last sixteen.

On average, during the ninety-two grand slam events in the period 1990–2012 (twenty-three years, four events per year), 53% of the top-sixteen men and 62% of the top-sixteen women reached the last sixteen. So there are *more* rather than fewer upsets for the men than for the women, contrary to what the previous section predicted. The only possible explanation is that the variation in strength $p_i - p_j$ is much smaller for the men than for the women, and this corresponds to casual observation.

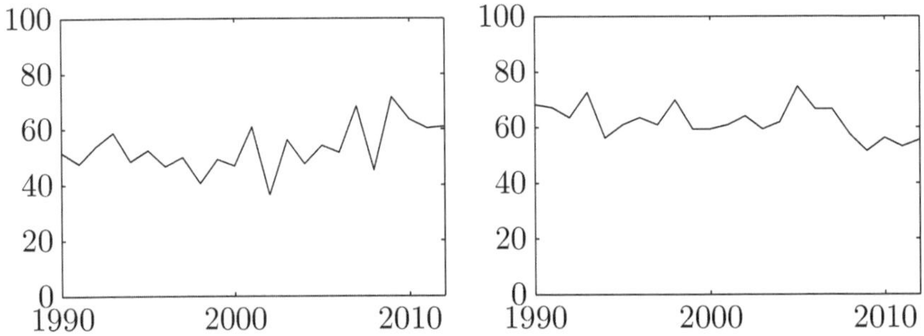

Figure 2.5: *Percentage of top-sixteen seeds reaching last sixteen, 1990–2012 (men left, women right)*

Have the men always experienced more upsets? Figure 2.5 plots the development of the percentages over time. We see an increase (fewer upsets) for the men and a decrease (more upsets) for the women in the last few years. If we consider only the years 2009–2012 and average over these four years and the four grand slams, then 64% of the top-sixteen men and 54% of the top-sixteen women reached the last sixteen. This comes closer to what we would expect based on the difference between best-of-three and best-of-five, and thus shows indirectly that the variation in strength has become larger for the men or smaller for the women or both. We return to this issue in Chapter 7 when we discuss hypothesis 12.

Long matches: Isner-Mahut 2010

Long matches occur from time to time. The 1969 first-round Wimbledon match between Pancho Gonzales and Charlie Pasarell lasted 112 games, but this was before the introduction of the tiebreak. At the US Open tiebreaks are played in all sets, including the final set, but in the other three grand slam tournaments there is no tiebreak in the final set and hence long final sets can occur. Andy Roddick defeated Younes El Aynaoui at the 2003 Australian Open quarter final in a five-set thriller with a score of 21-19 in the final set (eighty-three games in total), but this was nothing compared to the extraordinary match in which John Isner defeated Nicolas Mahut in the first round of Wimbledon 2010 after eleven hours and five minutes (over three days), with a score of 6-4, 3-6, 6-7, 7-6, 70-68 (183 games, 980 points).

At the grand slams almost 20% of the men's matches go to five sets. At the three grand slams where a tiebreak is not played in the final set, the number of 'long' (lasting at least fourteen games) fifth sets is about 15%. Since there are 127 matches to be played in each tournament, we would expect $0.20 \times 0.15 \times 127$ (that is, about four) long matches every year in each of the Australian Open, Roland Garros, and Wimbledon. Interestingly, there are also (on average) about four long matches in each women's tournament. This is because there are more final (third) sets, about 30%, but fewer of these are long, about 10%.

For a statistician to investigate an extreme event such as the Isner-Mahut match, the difficulty lies in the fact that very few observations are available. Very long matches are very rare. In such a situation we need more theory and more structure to produce credible results. Below we provide a mathematical analysis with a minimum of empirical observations. Still, the analysis allows us to make statements about how special the Isner-Mahut match was.

In calculating the probability of a long match we shall assume that hypothesis 1 holds, so that the points are iid. In addition, we shall assume that both players $\mathcal{I}$ and $\mathcal{J}$ have the same probability $p_i = p_j$ of winning their service point. Hence, they are equally strong. This is the situation where long matches can occur and hence the case of interest to us. Under these two assumptions, both players have the same probability, denoted by g, to win a service game.

Next we observe that for a long match to occur two 'knots' must be passed. The players must reach 2-2 in sets and in the final set they must reach 5-5 in games. There is no other way that a long match can develop. As a result, we can do the calculations in three steps.

First, the probability of reaching 2-2 in sets equals 3/8. This is because we have assumed that $p_i = p_j$, so that the probability that

$\mathcal{I}$ wins a set is 1/2. There are six possibilities to reach 2-2 in sets:

$$\mathcal{IIJJ},\quad \mathcal{IJIJ},\quad \mathcal{IJJI},\quad \mathcal{JIIJ},\quad \mathcal{JIJI},\quad \mathcal{JJII}.$$

Each of these has a probability of 1/16. Hence the probability of reaching 2-2 in sets is $6/16 = 3/8$. This is an example of the binomial probability distribution.

Second, we can use the binomial distribution to obtain the probability of reaching the score 5-5 in a set when both players have probability g to win their service game. This is more difficult and we simply denote this probability by $\ell(5,5)$ (ℓ for likelihood). The higher g, the more likely 5-5 becomes.

Third, we compute the probability that — given a 5-5 score in the final set — the set is decided with a score of a-b (a games for $\mathcal{I}$ and b games for $\mathcal{J}$). Multiplying this probability with $\ell(5,5)$ gives $\ell(a,b)$, the probability that the final set reaches the score a-b. The probability that two players of equal strength have to compete until a-b in the final set is then equal to $(3/8) \times \ell(a,b)$. Obviously, b-a has the same probability.

What is the probability of a match as extreme as the Isner-Mahut match, that is, a match ending 70-68 in the final set? It is

$$2 \times (3/8) \times \ell(70,68).$$

More important is the probability of a match *at least as* extreme as the Isner-Mahut match. This probability is given by

$$2 \times (3/8) \times \big(\ell(70,68) + \ell(71,69) + \ell(72,70) + \dots\big),$$

which depends only on the game-winning probability g which, in turn, depends only on the common point probability p. Table 2.2 presents the implied probabilities for each of four values of p.

Point prob., p	60	70	80	90
Game prob., g	73.6	90.1	97.8	99.9
Prob. of long match	3.6×10^{-13}	7.1×10^{-5}	2.0	30.8

Table 2.2: *Probability (in %) of a match as long as or longer than Isner-Mahut*

On average, the probability of winning a point of service is about 67% at Wimbledon (for the men; for the women it is 59%). At the Isner-Mahut match John Isner won 76.2% of the points on his service and Nicolas Mahut 78.7%. Hence the assumption $p_i = p_j$ is not unreasonable. The probability of two players of about equal strength with such high values of p playing against each other is small. But if they meet, then the probability of a long match is not that small: for $p = 80\%$ the probability of a match with a fifth set as long or longer than Isner-Mahut is 2%.

Rule changes: the no-ad rule

Rule changes are regularly discussed in order to make matches more exciting and attractive to watch. Rules to make matches shorter include the tiebreak (now standard), but also more controversial proposals such as the 'match tiebreak', the 'short set', and the 'no-ad' rule. These alternatives have long been used by amateurs and in selected competitions, such as World Team Tennis and Intercollegiate Tennis, but they were only formally recognized in the ITF Rules of Tennis from 2002 onwards.

A match tiebreak (also called supertiebreak) replaces the final set. Instead of playing a final set, one plays an extra long tiebreak: a match tiebreak. The winner of a match tiebreak is the player who reaches ten points first (rather than seven as in the ordinary tiebreak) with at least a two-point advantage. If a two-point advantage is not reached then the match tiebreak continues until it is.

A short set is a four-game set where the first to win at least four games with a margin of at least two games wins the set. A tiebreak is played at four-games-all. This rule is sometimes proposed together with playing best-of-five sets in all tournaments for both men and women.

The no-ad rule replaces the deuce system. The first player to win four points wins the game, even if the score is deuce. At deuce only one more point is played and the receiver has the choice of deuce court (right) or ad court (left).

The match tiebreak and the no-ad rule are currently being used for doubles matches in ATP and WTA tournaments, but not in the grand slam tournaments. These rules lead to shorter and more predictable (in terms of time) doubles matches — in statistical terms, they lead to a lower mean and a lower variance in match length. Although the doubles players were initially opposed, they were persuaded with the promise of more doubles matches on the principal courts. It was also hoped that more of the top singles players would play doubles with this format.

An experiment to play best-of-five sets for both men and women but with 'short sets' was conducted in the late 1990s at lower ITF tournaments, but the players' response was negative and the trial discontinued.

Let us consider the no-ad rule. If the traditional scoring system at deuce would be replaced by the no-ad rule, so that only one deciding point is played at deuce, then the probability g_i on page 15 that $\mathcal{I}$ wins the game would change to

$$\tilde{g}_i = p_i^4(-20p_i^3 + 70p_i^2 - 84p_i + 35).$$

It is easy to see that for $p_i > 0.5$ (the most common case), we have $\tilde{g}_i < g_i$, so that more service breaks will occur. The largest discrepancy occurs at $p_i = 0.65$, where $g_i = 0.83$ and $\tilde{g}_i = 0.80$, and the probability of a break thus increases from 17% to 20%.

Abolishing the second service

Rules have also been proposed to reduce the service dominance. One could allow larger balls, which would imply better visibility on television and slower movement through the air. Such balls have in fact been produced, but they are seldom used. The most obvious rule change, however, would be to abolish the second service. There is no particular reason why the server should have two possibilities to serve and there is no other sport where such a rule exists. Bill Tilden, a ten-time grand slam winner, speculated in 1920 that the second service would eventually be abolished, but so far he has been proven wrong. What would be the consequence of abolishing the two-service rule?

Most people — when asked this question — respond that with only one service a player would serve 'somewhere in-between' his or her current first and second service. But this would not be a good strategy. The correct strategy is to simply forget about the current first service and always use the current second service. It is easy to see why. A player with only one service is equivalent to a player with two services having faulted the first service. So, with only one service, a player should use his or her current second service, not something in-between. In the language of game theory (a branch of mathematics), the current situation (with two services) has a subgame (the second service) and in a subgame perfect equilibrium one plays the second service as in the equilibrium of the game in which only one service is available. Hence, the proposed change to one service actually amounts to abolishing the *first* service.

To see the effect of this proposal we consider the probability of winning a point on service. Under the current two-service rule this probability is quantified by the relative frequencies in Table 2.1. Data under the proposed one-service rule are not available, because this rule is currently not applied. However, the above game-theoretical argument implies that we can simply take the relative frequency of winning a point on the second service, and these data are available.

	Men		Women	
	2 serves	1 serve	2 serves	1 serve
Australian Open	62.2	49.9	55.4	44.8
Roland Garros	62.4	50.6	55.7	44.9
Wimbledon	66.7	51.9	59.3	46.9
US Open	63.6	51.4	55.7	44.8

Table 2.3: *Estimated probability of winning a point on service under the two- and one-service rules, respectively, 2010 data*

Table 2.3 shows that at Wimbledon the probability of winning a point on service for the men would decrease from 66.7% to 51.9%, still a service advantage, but much smaller than before. For the women, it would decrease from 59.3% to 46.9%. Here the service advantage would turn into a disadvantage. The idea of a service disadvantage may seem strange to tennis fans. But there are many sports (such as volleyball) where this is the case. Both new percentages are closer to 50% than before. Hence there will be more breaks and the probability of very long matches will be reduced.

One might argue that, if the one-service rule is implemented, players will spend more time practicing that service, so that the winning probability may increase. On the other hand, players will also adjust their training on returning the single service, thus reducing the service-winning probability. The overall training effect can go either way, and we ignore it.

Other consequences of rule changes, particularly on the optimal service strategy, will be examined in Chapter 9.

Further reading

The assumption that points are iid has been the subject of several studies. Klaassen and Magnus (2001) show that iid is rejected but that deviations from iid are small, particularly for top players, so that imposing iid will still provide a good approximation in many cases (see also Chapter 10). Newton and Aslam (2006) confirm the latter result, even when relatively strong non-iid effects are introduced in their simulations. But is it optimal to play points 'as they come' and play (close to) iid? Walker *et al.* (2011) derive that it is indeed optimal to ignore the score, that is, play according to an identical distribution.

Under iid one can calculate the probability of winning a game, set, tiebreak, or match, as *Richard* does. This has a long history, beginning with Kemeny and Snell (1960, pp. 161–167), and including Hsi and Burych (1971), Morris (1977) (who also discusses the magnification effect), Riddle (1988, 1989) (who also proves that serving first in a set does not matter under iid), and Newton and Keller (2005).

These approaches have been used to study the tennis scoring system and its effect on the probability of winning a match, naturally leading to some new proposals for rule changes. Pollard (1983) shows that the better player wins more often in a classical set (without tiebreak) than in a tiebreak set. Miles (1984) argues that in top men's tennis the proportion of service points won is so high that many points have to be played to identify the better player. It would be more efficient (and also lead to shorter matches, easier for television scheduling) to begin each game at 0-30 rather than at 0-0. Another innovative idea is due to Pollard and Noble (2004), who suggest the '50-40 game': the server must still win four points in order to win the game (reach score 50), but the receiver requires only three points in order to win it (40). This leads to a more predictable match duration in a match between strong servers. For an analysis of the consequences of abolishing the second service, see Klaassen and Magnus (2000).

Regarding upsets, Boulier and Stekler (1999) report that in grand slam tournaments during 1985–1995, 53% of seeded male players and 63% of seeded female players reached the last sixteen. The 1990–2012 results in this book (53% and 62%, respectively) are similar. Magnus and Klaassen (1999c) find that for the men a seeded player beats a non-seed somewhat less frequently than for the women. All results suggest that men's tennis is more competitive than women's tennis. A more powerful analysis confirming this statement will be provided when we test hypothesis 12 in Chapter 7.

3 Forecasting

The use of statistics has become increasingly popular in sports. Television broadcasts inform us about the percentage of ball possession in football, the number of home runs in baseball, and the number of aces and double faults in tennis. All these statistics provide some insight into the question which player or team performs particularly well in a match, and therefore also (indirectly) into the question who is more likely to win. Surprisingly, however, a direct estimate of the probability that a player or a team wins the match is not shown.

In this chapter we provide such a direct estimate for tennis. We forecast the winner of a match, not only at the beginning of the match, but also while the match is unfolding; in fact, at each point. The forecast is produced within one second after each point, and the resulting profile of winning probabilities provides a quick overview of the match developments so far and a direct forecast of who will win the match. The quality of the profile is confirmed by the probabilities implied by in-play betting odds.

The profile provides valuable insights for fans and commentators watching a tennis match. The score itself, obviously an important ingredient in forecasting the outcome, tells us which player is leading, but this does not imply that this player is also likely to win the match: a top player may still be the favorite after losing the first set. Moreover, a score of 5-5 can result after 4-4, but also after 5-0, so that the score provides only partial information on the development of the match so far. Summary statistics do not fill these voids, but a graph of winning probabilities does. The profile should thus prove to be a powerful tool for commentators and viewers.

Forecasting with *Richard*

Richard, the computer program introduced in the previous chapter, will be an essential ingredient in our forecasting exercise. In a match between two players $\mathcal{I}$ and $\mathcal{J}$, *Richard* calculates the probability that $\mathcal{I}$ wins the match. To explain the procedure, let p_i be the probability that $\mathcal{I}$ wins a point on service and let p_j be the probability that $\mathcal{J}$ wins a point on service, as in the previous chapter. Then, under hypothesis 1 that points are independent and identically distributed, the match probability m_i at the beginning of the match depends only on the two point probabilities (p_i and p_j) and the type of tournament (best-of-three sets or best-of-five sets, tiebreak in final set or not).

This is at the beginning of a match. *During* the match we can also employ *Richard* to calculate the probability m_i that $\mathcal{I}$ will win at the beginning of the point under consideration. The match probability m_i then depends, in addition, on the current score and the current server.

In a specific match between $\mathcal{I}$ and $\mathcal{J}$ we know everything except the point probabilities p_i and p_j. It is difficult to obtain credible values for them directly. For example, what is the probability that Roger Federer will win a point on service against Rafael Nadal when they have not yet played a point? Moreover, values that differ just slightly from the true p_i and p_j can lead to a completely different match probability m_i due to the magnification effect, as shown in Chapter 2. The difference $p_i - p_j$, in particular, has a major impact on m_i due to the steepness of the graphs in Figure 2.4. Obtaining sufficiently precise values of p_i and p_j is nearly impossible, at least when we attempt to estimate them directly.

Our solution to this problem is to exploit the fact that it is relatively easy to obtain a value for the match probability m_i at the beginning of a match. We combine this with the theory of Chapter 2 to determine p_i and p_j indirectly. More specifically, we know that from p_i and p_j, or equivalently from $p_i - p_j$ and $p_i + p_j$, *Richard* calculates m_i at the beginning of a match. So, knowing m_i and $p_i + p_j$, we can calculate $p_i - p_j$ by 'inverting' *Richard*. In terms of Figure 2.4 this inversion implies going from a value of m_i on the vertical axis via the S-shaped curve implied by $p_i + p_j$ to a unique value of $p_i - p_j$ on the horizontal axis. The inversion automatically circumvents the problem that small deviations in $p_i - p_j$ have a major impact on m_i, because we start from m_i, not from $p_i - p_j$. We thus transform the initial problem of finding p_i and p_j into two (easier) problems: finding m_i at the beginning of the match and finding $p_i + p_j$. From m_i and $p_i + p_j$ we first derive $p_i - p_j$, and then p_i and p_j themselves. How do we know m_i at the beginning of a match and how do we determine $p_i + p_j$?

There are several ways to estimate m_i at the beginning of a match. One could use the rankings of the two players. This works reasonably well in practice but it has the disadvantage that the court surface (grass, clay) is not taken into account and also that recent information such as a minor injury is ignored.

A different way, incorporating *all* publicly available information, is to use betting odds. The fractional odds in favor of an event are defined as the ratio of the probability that the event will happen to the probability that it will not happen. For example, the fractional odds that a randomly chosen day of the week is a Sunday are one to six, written as 1:6. This system is favored in the United Kingdom and Ireland and is common in horse racing, but it is a little counterintuitive.

More intuitive are the decimal odds, which are simply the inverse probability. The decimal odds that a randomly chosen day of the week is a Sunday are seven to one, written as 7.0. If the decimal odds of $\mathcal{I}$ defeating $\mathcal{J}$ are 1.5, then m_i is the inverse, so $m_i = 2/3$. In fact, the transformation from betting odds to match probability m_i is a little more complex, because the decimal odds do not represent the true inverse probabilities as perceived by the bookmaker, but are the amounts that the bookmaker will pay out on winning bets. In formulating his odds the bookmaker will include a profit margin, the so-called overround on his book, for which we will account when computing m_i.

We also need a value for $p_i + p_j$. This too can be achieved in several ways. Using the rankings of the two players and a statistical model is one option. But a simpler alternative is to just set $p_i + p_j = 2\bar{p}$, where $\bar{p}$ denotes the tournament average of the players' point-winning probability p in a representative year. This average can easily be estimated by combining the summary statistics of all matches in that year, and Table 2.1 presents the values for the grand slam tournaments in 2010. We shall use this simpler method in the analysis below.

The reason that the simple method works in practice is that, at many points, m_i does not much depend on $p_i + p_j$, as we saw in the previous chapter. Still, there may be points where $p_i + p_j$ matters. For two players $\mathcal{I}$ and $\mathcal{J}$ in a specific match, we should perform a sensitivity analysis to check whether the match profile changes when the estimate of $p_i + p_j$ is changed. We shall do so below and conclude that the influence on the match profile is small and that the proposed simple method thus accurately estimates the match probabilities.

Summarizing, we calculate the probability m_i that $\mathcal{I}$ will win the match against $\mathcal{J}$ as follows. First we compute m_i at the beginning of the match from betting odds. Then we compute $\bar{p}$ as the tournament average in a comparable year, and set $p_i + p_j = 2\bar{p}$. Given m_i and $p_i + p_j$ we then call on *Richard* to produce $p_i - p_j$. Once we have $p_i + p_j$ and $p_i - p_j$, we also have p_i and p_j separately, so that we can compute m_i, not only at the beginning of the match, but at each point as the match unfolds. *Richard* calculates very quickly, and the updated m_i is available within a second after completion of the point.

Federer-Nadal, Wimbledon final 2008

We illustrate the forecasting procedure by analyzing three famous matches. First, the 2008 Wimbledon final between Roger Federer (seeded 1) and Rafael Nadal (seeded 2). Federer had won the five previous finals at Wimbledon and had he won this time he would have become the first man since William Renshaw (1881–1886) to win a sixth consecutive championship at The All England Lawn Tennis and Croquet Club. He came close, but in the end he lost 4-6, 4-6, 7-6, 7-6, 7-9 in arguably the best match ever played at Wimbledon. It was Nadal's first Wimbledon title, having lost to Federer in the 2006 final in four sets, and in the 2007 final in five.

At the beginning of the match Federer was the favorite. Let us call him player $\mathcal{I}$ and Nadal player $\mathcal{J}$. Averaging over five betting sites, we obtain decimal betting odds of $b_i = 1.691$ (Federer) and $b_j = 2.198$ (Nadal). Direct calculation of the match-winning probabilities would give $1/b_i = 0.591$ and $1/b_j = 0.455$, respectively. These two 'probabilities' add up to 1.046, hence more

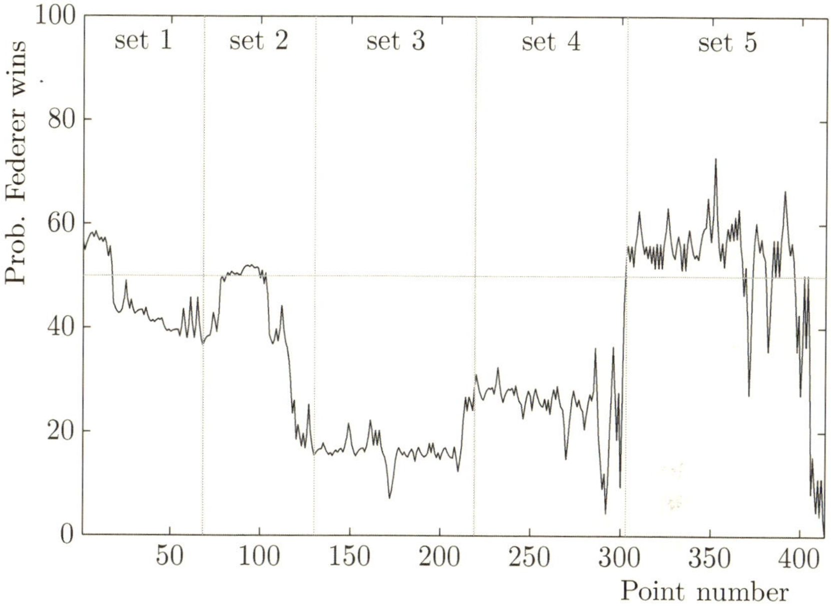

Figure 3.1: *Federer-Nadal profile, Wimbledon 2008 ($\bar{p} = 66.7\%$)*

than one because of the overround. Correcting for this we obtain $m_i = 0.591/1.046 = 0.565$ and $m_j = 0.455/1.046 = 0.435$. Hence we estimate the initial probability that Federer wins as $m_i = 56.5\%$.

From Table 2.1 we find $\bar{p} = 66.7\%$. From $p_i + p_j = 2\bar{p} = 1.334$ and $m_i = 0.565$ we call on (inverted) *Richard* to calculate $p_i - p_j = 0.011$. This then gives $p_i = 67.2\%$ (Federer) and $p_j = 66.2\%$ (Nadal) from which the match profile can be computed.

The match profile is presented in Figure 3.1 from Federer's point of view. At the beginning of the match Federer's probability of winning was 56.5%. After losing the first two sets, this probability had dropped to 15.3%, but when Federer won the third-set tiebreak his probability of winning the match went up again, to 28.7%. Big swings occurred at the end of the fourth set. Nadal led 5-2 in the tiebreak, then double-faulted to 5-3. At 7-6 and again at 8-7, Nadal had matchpoints (the second time on his own service). Eventually Federer won the tiebreak 10-8.

No man had come back to win a Wimbledon final after losing the first two sets since Henri Cochet defeated Jean Borotra in 1927, but at 2-2 in sets Federer was still favorite to win with probability 53.8%, smaller than the original 56.5% (because of the magnification effect) but larger than 50%. In the final set, Federer was 4-3 ahead and had breakpoint on Nadal's service. His probability of winning the match was then 72.9%, the highest of the match. But Nadal held serve. Two games later, at 5-4, Federer needed only two points for the championship and his probability of winning was 62.9%. Again Nadal held service. At 5-5 Federer was two breakpoints down. Commentators were quick to point out that Federer's disappointment of missing the earlier chances had affected the momentum in Nadal's favor. Whether the commentators were right or wrong will be considered in Chapter 12, where we ask the general question of whether missed breakpoint(s) in one game lead to a larger probability of being broken in the next (hypothesis 22). Federer's probability dropped to 27.1%, but he managed to save the breakpoints. Both players continued to hold service until 7-7. Then Federer lost his service and Nadal served for the match to win 9-7 in the final set. After the last point the probability that Federer wins is obviously zero.

The profile provides a graphical illustration of the changing fortunes during the match, not only qualitatively but also quantitatively. A comparison with the 50% line makes it immediately clear who the current favorite is, and also by how much. Moreover, the graph can be produced while the match is in progress — we don't have to wait until the match is completed.

Effect of smaller $\bar{p}$

We already mentioned a possibly weak link in the theory underlying the graph, namely the determination of $p_i + p_j$. In computing the profile we used the tournament average $\bar{p}$ in a representative year, and set $p_i + p_j = 2\bar{p}$. This is simple, perhaps too simple. What is the effect when we refine the estimate of $p_i + p_j$?

At the 2006 Wimbledon final between Roger Federer and Rafael Nadal they realized $p_i + p_j = 1.320$, and at the 2007 final 1.339. (At the 2008 Wimbledon final between Federer and Nadal they realized $p_i + p_j = 1.337$, but of course we only knew this at the end of the match, so we could not use it during the match.) This corresponds to taking $\bar{p} = 66.0\%$ in 2006, 66.9% in 2007, and 66.8% in 2008.

We used $\bar{p} = 66.7\%$, based on Wimbledon 2010, and the difference between using this average and using any of the other three averages is negligible in the graph. But suppose that the match under consideration seriously deviates from the tournament average, so that we should have taken $\bar{p} = 60\%$ or 70% instead of 66.7%. What would be the effect on the profile?

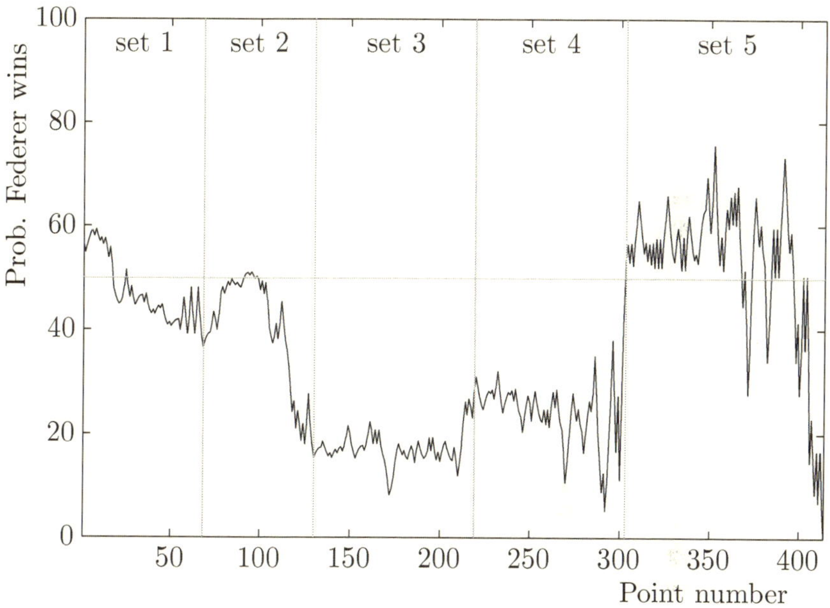

Figure 3.2: *Federer-Nadal profile, Wimbledon 2008 ($\bar{p} = 60\%$)*

In Figure 3.2 we consider the case $\bar{p} = 60\%$, which is much lower than the 66.7% used before. With $m_i = 0.565$ and $p_i + p_j = 2\bar{p} = 1.2$ *Richard* produces $p_i - p_j = 0.010$, so that $p_i = 60.5\%$ (Federer) and $p_j = 59.5\%$ (Nadal). Qualitatively the new profile looks the same as the previous one, and even quantitatively the difference is small. For example, after losing the first two sets, Federer's probability m_i of winning dropped from 56.5% at the beginning of the match to 15.3% (previously also 15.3%). At 6-6 in the fourth set with Nadal leading 5-2 in the tiebreak, the probability m_i was 5.1% (previously 4.2%). And at the beginning of the final set we find $m_i = 53.6\%$ (previously 53.8%).

These numbers confirm what we found in Chapter 2. The profile is not sensitive to the specification of $p_i + p_j$.

Kim Clijsters defeats Venus Williams, US Open 2010

Venus Williams (seven grand slam singles titles) won both Wimbledon and the US Open in 2000 and 2001, and she won Wimbledon again in 2005, 2007, and 2008. Kim Clijsters won the US Open in 2005. The next three years she did not participate, but in 2009 she received a wildcard and, only one month after her return to the professional tour, she won again, defeating Caroline Wozniacki in the final. In 2010, Clijsters (seeded 2) met Williams (seeded 3) in the semi-final, and defeated her in a memorable three-set match: 4-6, 7-6, 6-4.

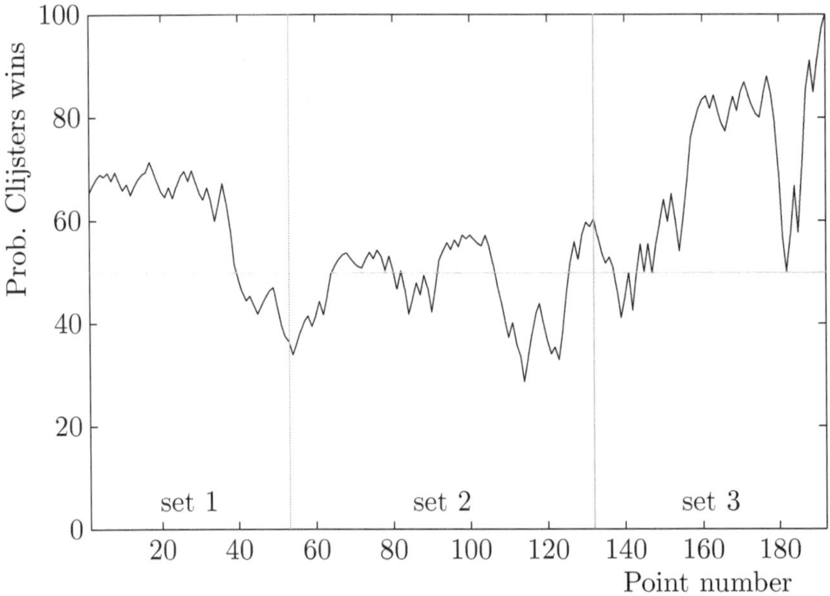

Figure 3.3: *Clijsters-Williams profile, US Open 2010 ($\bar{p} = 55.7\%$)*

Clijsters, the defending champion, was the favorite at the beginning of the match. Averaging over five betting sites, we find decimal odds of $b_i = 1.452$ (Clijsters) and $b_j = 2.732$ (Williams).

Similar to the calculation for Federer-Nadal on pages 36–37, we derive $m_i = 0.653$ as the initial winning probability for Clijsters. Together with $\bar{p} = 55.7\%$ from Table 2.1 this gives $p_i = 57.2\%$ for Kim Clijsters and $p_j = 54.2\%$ for Venus Williams.

The match profile is presented in Figure 3.3 from Clijsters' perspective. Kim Clijsters served first and her winning probability at the beginning of the match was 65.3%. At 3-3 Clijsters lost her service game, and at that point m_i had dropped to 51.5%. After losing the first set 4-6, m_i had dropped further to 36.4%.

In the second set Clijsters won the first three games, but at 5-2 she lost three consecutive games. The best score from Williams' perspective occurred at 5-5, when Clijsters, serving, was 0-15 behind and her probability of winning had dropped to 28.7%. But Clijsters held her service game, as did Williams thereafter. The tiebreak was convincingly won 7-2 by Clijsters.

At the beginning of the final set m_i was 60.3%. Williams served first in the set and won her first service game. In the second game Clijsters was 0-30 and 30-40 behind, making Williams the favorite to win the match, but Clijsters held serve. At 1-1 Williams lost her service game. At 4-2, m_i had increased to 86.8%, but when Williams won the next two games m_i dropped again to 56.7%. When Williams won the first point on her service in the next game, the match was completely balanced. But Williams lost her service and Clijsters served successfully for the match, letting m_i increase rapidly from 50.1% to 100%. In the final, Kim Clijsters defeated Vera Zvonareva in less than one hour.

The swings in the match are clearly visible, in particular the swing at the end of the first set, the swing in the second set from 5-2 via 5-5 to 7-6, and the big swings in the final set. The graph provides insights that a score cannot reveal.

Effect of larger $\bar{p}$

Again, we ask about the effect on the profile when we assume a different value for $p_i + p_j$. Comparing with two other hard court matches, we find that at the 2009 US Open (fourth round), Kim Clijsters and Venus Williams realized $p_i + p_j = 1.145$; and at the 2010 final in Miami, 1.215. This corresponds to taking $\bar{p} = 57.2\%$ and 60.8%, respectively, while we used $\bar{p} = 55.7\%$. Below we pro-

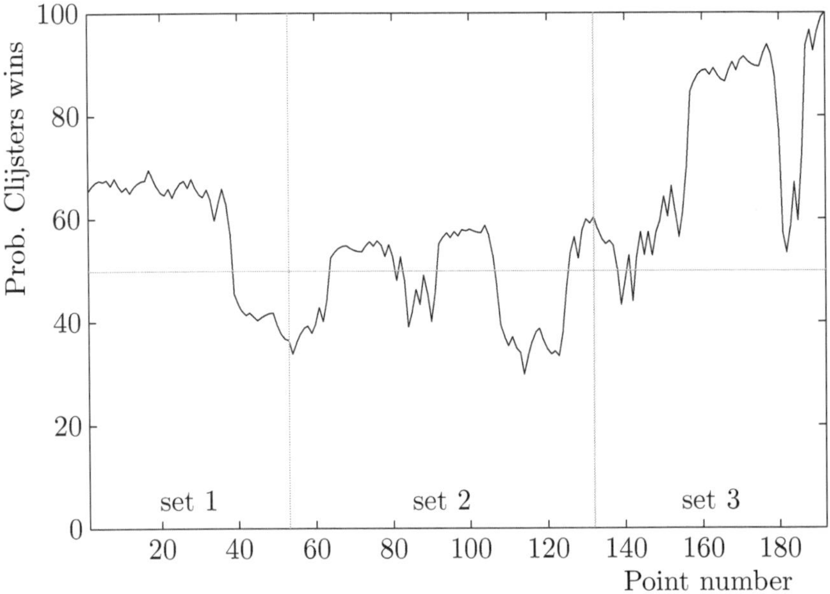

Figure 3.4: *Clijsters-Williams profile, US Open 2010* ($\bar{p} = 65\%$)

vide the profile for an even higher value of $\bar{p}$, namely 65%.

What is the effect on the profile? Of course, there is a quantitative effect, but it is very small even for such a large deviation, and the shape of Figure 3.4 is almost identical to the shape of Figure 3.3. Again we conclude that the profile is not sensitive to the specification of $p_i + p_j$.

Djokovic-Nadal, Australian Open 2012

Finally, from the Australian Open, we look at the profile of the 2012 final, where Novak Djokovic (seeded 1) defeated Rafael Nadal (seeded 2) in five sets: 5-7, 6-4, 6-2, 6-7, 7-5. It was the longest grand slam final ever: five hours and fifty-three minutes.

Djokovic was the favorite. Averaging over seven betting sites, we estimate the decimal odds as $b_i = 1.676$ for Djokovic and $b_j = 2.257$ for Nadal. This yields $m_i = 0.574$ as the initial winning probability for Djokovic. Together with $\bar{p} = 62.2\%$ from Table 2.1 we get $p_i = 62.8\%$ (Djokovic) and $p_j = 61.6\%$ (Nadal).

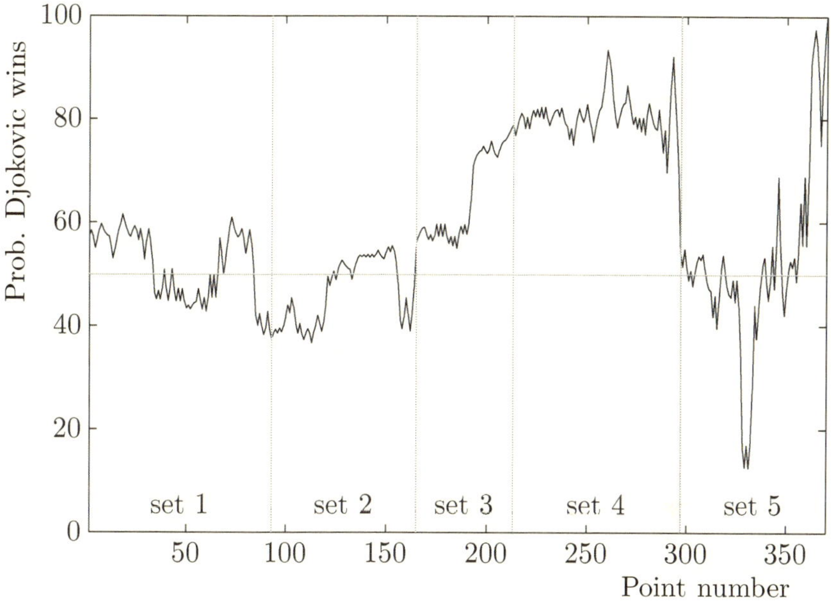

Figure 3.5: *Djokovic-Nadal profile, Australian Open 2012* ($\bar{p} = 62.2\%$)

The match profile is presented in Figure 3.5 from the point of view of Djokovic. Djokovic started the match with a winning probability of 57.4%. At 4-2 for Nadal in the first set, m_i had dropped to 43.3%, increasing to 56.8% at 5-5. After Nadal won the first set 7-5, m_i had dropped to 37.4%, and Nadal was favorite to win the match. After Djokovic won the second set his probability of winning the match increased to 56.0% and after winning the third set to 78.9%.

At the beginning of the tiebreak in the fourth set, Djokovic still had a winning probability of 78.0%, and at 5-3 in the tiebreak (two points from victory) even 92.2%. But Nadal won the tiebreak 7-5 and m_i dropped sharply to 54.2%.

In the final set Nadal broke Djokovic in the sixth game. At that point, 4-2 for Nadal in the final set, m_i has dropped to 16.1%, and at 30-15 for Nadal even to 12.5%. This was the point in the match with the highest probability for Nadal to win: 87.5%. Then, Djokovic broke Nadal's service, won four of the next five games, and the match.

In-play betting

Our forecasting profiles are based on *Richard*, a computer program, which in turn is based on a model. The only assumption in this model is that the service points of each player in a match are iid. The three match examples suggest that the profiles provide a realistic picture of the matches involved, and hence provide indirect evidence that the model makes sense. But what does 'the market' tell us about the quality of our profiles?

The emergence of in-play tennis betting markets allows us to compare our *Richard*-based forecasts to the forecasts implied by betting odds. We focus on one match, the Federer-Nadal Wimbledon 2008 final. To compare the market forecasts to our profiles we need a market assessment of the probability that Federer wins the match at each point of the match.

The market's view is computed from realized trades at Betfair, an online betting company launched in 2000. Betfair is an exchange where one gambler bets against another gambler (hence not against a bookmaker). Betfair acts as an intermediary and earns commission on the profit of a winning gambler. This implies that there is no profit margin in the odds themselves (overround) as in the case of a bookmaker.

From the matched odds one can directly infer the implied probability of an event. More specifically, the probability that Federer wins the match is the inverse of the (decimal) odds on Federer to win. An important match such as a Wimbledon final attracts a large amount of betting. The value of matched bets was six million pounds at the beginning of the match, and increased to fifty million pounds at the end. The market is therefore big enough to consider the implied probabilities as the market's view on the evolving probabilities that Federer wins the match.

From a few hours before the match until the end of the match the probability is available, not point by point but second by second. Our model-based probability is measured at each point, not at each second. In order to provide a comparison we transformed the Betfair data to our format, thus obtaining the market's winning probability at the beginning of each point.

Figure 3.6 replicates our forecast profile from Figure 3.1 (dark line) and adds the Betfair probabilities (light line). The correspon-

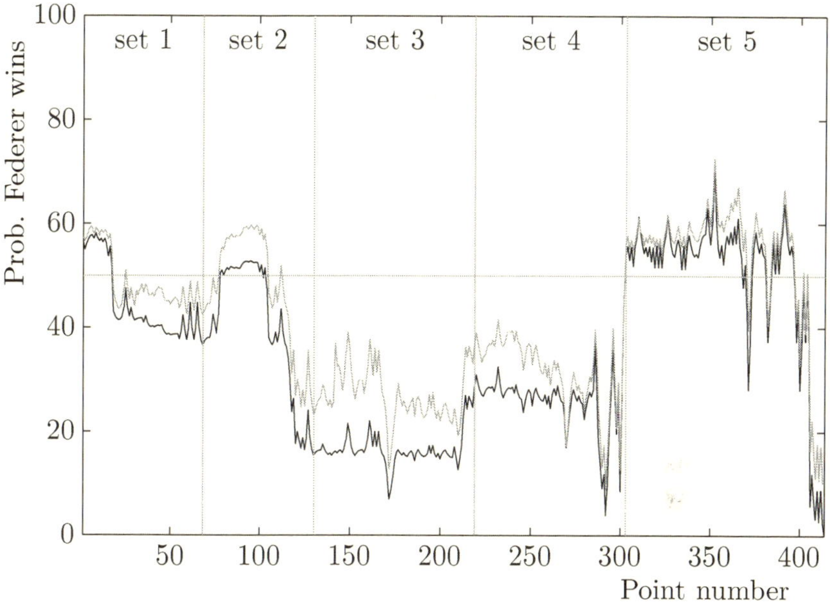

Figure 3.6: *Model (dark) and Betfair (light) profiles, Federer-Nadal at Wimbledon 2008*

dence is striking, in fact so striking that the only credible explanation is that Betfair traders and their computers are using our model or a model very similar to ours.

Let us take a closer view. At the beginning of the match the two probabilities virtually coincide, with our approach giving Federer 56.5% chance to win and Betfair 57.5%. This is not surprising, because our starting probability is derived from the bookmakers odds. As the match unfolds, some differences between the two graphs emerge, not so much in the movement from one point to the next, but rather in the level. From the middle of the first set until halfway through the fourth the *Richard* probabilities are about 10%-points below the Betfair probabilities. For example, at 0-1 in sets, 4-2 in games, and 30-30 in points on Federer's service (point 103), the probabilities are 50.6% and 58.1%, respectively. This is a large difference. In fact too large, given the starting probability of around 57%. It is difficult to explain why Federer's match-winning probability at this point would be higher than at the beginning of the match. Still, this is apparently what the market believed.

How to explain this discrepancy? Let us use to same starting point as Betfair, that is, 57.5% rather than 56.5%. This increases the probability at point 103 from 50.6% to 51.5%, a slight increase, but insufficient to explain such a large discrepancy. Recall from page 35 that, apart from the initial match probability, our method requires an estimate of the sum of the point-winning probabilities p_i and p_j of Federer and Nadal. However, Figure 3.2 shows that, even if we decrease or increase this sum substantially, it does not much affect the probability at point 103.

The only remaining explanation is that Betfair traders allow for deviations from the iid assumption. For example, they might think that the favorite will win more often than predicted under the iid assumption, if the favorite is behind. This raises an interesting question: can such deviations from iid be uncovered from the data? We shall come back to this question in Chapters 10–12, where we discuss and apply methods to test for deviations from iid.

Further reading

In-play forecasting of the winner of a match dates back to Magnus and Klaassen (1995, 1997). There we used the seeding list to estimate the initial match-winning probability (the starting point of the profile). Using betting odds, as implemented in this book, appears to be an improvement.

Several alternatives for estimating the initial probability exist, based on past performances of the players. Blackman and Casey (1980) use match scores, while Clarke and Dyte (2000) use world-ranking points. Klaassen and Magnus (2003a,b,c; 2008) employ the positions on the world ranking, although the user of the profile (say, the commentator) can adjust the estimate to account for his or her own specific knowledge, such as special abilities of the players on the court surface or health problems. McHale and Morton (2011) utilize games won and lost in past matches and the time passed since these matches were played to capture recent form.

Barnett (2006) describes in-play forecasting and how it has been used since Wimbledon 2003 for setting bookmaker prices for betting on the point score in completed games, that is, the server or receiver winning to 0, 15, 30, or deuce. Jackson (1994) predicted a booming sports-betting market, and the introduction of Betfair has validated this prediction; see Davies *et al.* (2005) for details on Betfair. The resulting forecasts are now also used for statistical analysis outside the betting scene. For example, Easton and Uylangco (2010) compare the predictions from our model to those derived from Betfair odds for forty-nine matches at the 2007 Australian Open and conclude that there is a very high correlation, again implying that the model predictions make sense.

4

Importance

Are all points equally important? Or are some points more important than others? If so, what are the 'big' points? Is 15-30 the most important point in a game? After all, if you (the server) win the point, then the score is 30-30 and you will probably win the game. But if you lose the point, it is 15-40 and you will probably lose the game. Is the seventh game the most important in a set, as one often hears? Which are the most important points in a match? All these questions relate to the concept of 'importance'.

What is importance?

Intuitively, a point is important within a game if winning or losing the point has a large impact on winning the game. Likewise, a game is important within a set if winning or losing the game has a large impact on winning the set. We make this intuition explicit starting from the simplest case: a point in a game. Then, we extend the concept step by step, so that in the end we will be able to compute the importance of each point in the match.

Consider a game where player $\mathcal{I}$ is serving against player $\mathcal{J}$. Within this game consider the point a-b, where a and b denote the scores of $\mathcal{I}$ and $\mathcal{J}$, respectively. The scores a and b can be 0, 15, 30, 40, and advantage (Ad). In our formulas this counting system is not convenient, so we shall often count points as 0, 1, 2, 3, 4, where the number 4 only appears in 4-3 or 3-4, advantage server or receiver. We use $g_i(a, b)$ to denote the probability that $\mathcal{I}$ wins the game from point a-b. Thus $g_i(0, 0)$ is the probability at the beginning of the game, which we abbreviated to g_i in Chapter 3.

The importance $imp_{pg}(a, b)$ of the point a-b for winning a game depends on two probabilities: $g_i(a + 1, b)$, the probability that the server wins the game given that he or she *wins* the point; and $g_i(a, b + 1)$, the probability that the server wins the game given that he or she *loses* the point. Importance should depend on these two probabilities and on nothing else. But how should it depend on these probabilities? In 1977 Carl Morris suggested the simplest relationship, namely the difference:

$$imp_{pg}(a, b) = g_i(a + 1, b) - g_i(a, b + 1),$$

and this definition has been generally adopted. Since our program *Richard* calculates the probability of winning a game (and set and match) from a given score, it can also calculate the difference between these winning probabilities. Hence, *Richard* can be used directly to calculate importance variables.

With this definition in mind, we consider the following hypothesis.

Hypothesis 3: *Every point (game, set) is equally important to both players.*

One often hears: 'This is an important point, especially for $\mathcal{I}$', where $\mathcal{I}$ is typically the player who is behind. If $\mathcal{I}$ is down 15-40 in a game, is it perhaps more important for $\mathcal{I}$ to save this breakpoint than for $\mathcal{J}$ to convert it? No, it is not. We can prove this formally by writing

$$\begin{aligned} g_i(a + 1, b) - g_i(a, b + 1) &= (1 - g_j(a + 1, b)) - (1 - g_j(a, b + 1)) \\ &= g_j(a, b + 1) - g_j(a + 1, b), \end{aligned}$$

which says that the importance of point a-b is the same for $\mathcal{I}$ and $\mathcal{J}$. It is therefore *not* the case that points are more important for the player who is behind.

Big points in a game

Now that we know how to obtain the importance of a point in a game, we may ask: what are the important points, the 'big' points, in a game? Assuming still that $\mathcal{I}$ is serving with probability p_i of winning a service point, we can calculate the importance of 40-30 as one minus the probability that $\mathcal{I}$ wins the game from deuce, and the importance of 30-40 as the probability that $\mathcal{I}$ wins the game from deuce:

$$imp_{pg}(40,30) = \frac{(1-p_i)^2}{p_i^2 + (1-p_i)^2}, \quad imp_{pg}(30,40) = \frac{p_i^2}{p_i^2 + (1-p_i)^2}.$$

This shows that for $p_i > 0.5$, 30-40 (breakpoint) is more important than 40-30 (gamepoint), in line with intuition.

Given p_i the big points in a game can be quantified, and we illustrate this for $p_i = 64\%$ and $p_i = 57\%$, the average values in the 2010 grand slams for men and women, respectively.

		Point score receiver				
		0	15	30	40	Ad
	0	23	33	40	31	—
	15	17	30	45	49	—
Point score server	30	10	21	43	76	—
	40	3	9	24	43	76
	Ad	—	—	—	24	—

Table 4.1: *Importance of point in game,* $p_i = 64\%$ *(men)*

For the men, 30-40 is the most important point ($imp = 76\%$), much more important than 15-30 (45%); and 40-0 is the least important point. This, in fact, is true for any value of $p_i > 0.5$. The second most important point is not 15-30 but 15-40 ($imp = 49\%$), although this is not true for every value of $p_i > 0.5$. Note that 40-40 (deuce) and 30-30 are equivalent; and, similarly, that advantage server and 40-30, and advantage receiver and 30-40 are equivalent. Hence their importance must also be the same, and this is confirmed in Table 4.1.

For the women, the average probability of winning a point on service is $p_i = 57\%$, smaller than for the men but still larger than 50%. As a result, breakpoint (30-40) remains the most important point and 40-0 the least important. Table 4.2 shows that the second most important point is now 30-30 (or deuce) and not 15-40. In general, if $p_i \leq 61\%$ (as is typically the case for the women),

		Point score receiver				
		0	15	30	40	Ad
	0	29	35	33	21	—
	15	25	35	43	36	—
Point score server	30	17	30	48	64	—
	40	7	16	36	48	64
	Ad	—	—	—	36	—

Table 4.2: *Importance of point in game,* $p_i = 57\%$ *(women)*

then 30-30 is the second most important point, but if $p_i > 62\%$ (as is typically the case for the men), then 15-40 is second most important.

The closer p_i is to 50%, the less extreme are the importance values. When $p_i = 64\%$ the importance ranges from 3% to 76%; when $p_i = 57\%$ it ranges from 7% to 64%.

Big games in a set

The importance of games in a set is more complicated than of points in a game, because in a game we need only one probability but in a set we need two. In Tables 4.3 and 4.4 we assume that both players

		Game score j						
		0	1	2	3	4	5	6
	0	30	30	18	14	3	1	—
	1	30	33	33	17	12	1	—
	2	30	33	37	37	14	8	—
Game score i	3	14	32	37	42	42	9	—
	4	10	12	36	42	50	50	—
	5	1	6	8	41	50	50	50
	6	—	—	—	—	—	50	100

Table 4.3: *Importance of game in set,* $g_i = g_j =$ 81.3% *(men)*

are equally strong ($p_i = p_j$), so that they have the same chance of winning a point on service. This obviously implies that they have the same chance of holding their service games ($g_i = g_j$). In particular, we choose $g_i = g_j = 81.3\%$ (corresponding to $p_i = p_j = 64\%$ used earlier) for the men, and $g_i = g_j = 67.0\%$ (corresponding to $p_i = p_j = 57\%$) for the women.

There are two types of sets: those with a tiebreak and those without a tiebreak. The two tables concern sets with a tiebreak. Player $\mathcal{I}$ serves first in the set.

The most important game is clearly the tiebreak with *imp* = 100% followed by any other game at the end of a set, from 4-4 onwards. The score 0-0 is less important than 1-1, which is less important than 2-2, and so on, until 4-4. This is because at the beginning of a set there is more time to repair the loss of a game.

		Game score j						
		0	1	2	3	4	5	6
	0	26	26	20	16	7	2	—
	1	26	29	29	21	15	4	—
	2	25	29	33	33	20	11	—
Game score i	3	16	27	33	39	39	17	—
	4	11	15	30	39	50	50	—
	5	2	7	11	34	50	50	50
	6	—	—	—	—	—	50	100

Table 4.4: *Importance of game in set,* $g_i = g_j = 67.0\%$ *(women)*

For the women, Table 4.4 looks similar to Table 4.3, except that numbers on or close to the diagonal — such as 0-0, 3-2, or 3-4 — are smaller for the women than for the men, and that the opposite is true for numbers away from the diagonal — such as 4-1 and 2-4. This is because the probability of winning a service game is smaller for the women than for the men. As a consequence, if the score is close, say 1-1, then a break is more serious for the men than for the women, and hence close scores are more important for men than for women. On the other hand, if the score is not close, say at 4-1, then the woman trailing still has a chance, more than a man would have, and such a game is therefore more important for women than for men.

The vital seventh game

Another 'idée reçue' of commentators, and even of some players, is the mystery of the seventh game.

Hypothesis 4: *The seventh game is the most important game in the set.*

This fixation has existed for a long time, but it became popular with Dan Maskell, BBC tennis commentator for many years and the 'voice of Wimbledon' until he retired in 1991. He considered the seventh game 'all-important' and 'vital'. However, it appears from Tables 4.3 and 4.4 that there is nothing special about the seventh game. The score 3-3 is less important than 4-4, and 3-3 is equally important as 4-3. At 4-2, importance is lower than at 3-2 and 2-2.

Before rejecting the hypothesis, let us consider it more carefully. How would we define the importance of, say, the third game? The third game could be either at 2-0, 1-1, or 0-2. One of these three scores must occur in a set, but they do not occur with equal probabilities. The probabilities can be computed from the binomial probability function, which we already encountered on page 26. The score 2-0 occurs if player $\mathcal{I}$ holds service and breaks $\mathcal{J}$. Assuming again equal probabilities $g_i = g_j = 81.3\%$ of winning a service game for the men, the probability of 2-0 is equal to $g_i \times (1 - g_j) = 15\%$. The same probability applies to 0-2. However, for 1-1 we obtain

$$g_i \times g_j + (1 - g_i) \times (1 - g_j) = 70\%,$$

so that 1-1 is more likely than 2-0 and 0-2. If we are in the third game, we thus know the probability of each score, and for each score we know its importance imp_{gs} from Table 4.3. The importance of the third game is then given by the expected value of imp_{gs} over all three scores, that is,

$$imp_{gs}(3) = 0.15 \times 0.30 + 0.70 \times 0.33 + 0.15 \times 0.18 = 30\%.$$

Likewise, the ninth game can be either at 5-3, 4-4, or 3-5. If the ninth game occurs, the probability of 5-3 and 3-5 is 28% and the probability of 4-4 is 44%, from which we compute $imp_{gs}(9) = 36\%$.

But this is not the whole story. It only shows that the ninth game is more important than the third *if there is a ninth game.* However, the ninth game need not occur in a set, in contrast to the third. This matters, as the tiebreak exemplifies: conditional on its occurrence, the tiebreak is the most important game of all with $imp_{gs}(13) = 100\%$. Apparently the conditional concept imp_{gs} does not yet fully cover the idea of importance contained in hypothesis 4. We should account for the fact that some games occur less frequently than others.

Thus, let $\ell(n)$ be the probability that the nth game occurs. Clearly, $\ell(3) = 100\%$, but $\ell(9)$ is smaller than 100%: $\ell(9) = 84\%$. The unconditional importance that we require for the seventh-game hypothesis is then given by

$$imp_{gs}(3) \times \ell(3) = 0.30 \times 1.00 = 30\%,$$
$$imp_{gs}(9) \times \ell(9) = 0.36 \times 0.84 = 30\%.$$

	Game number n									
	1	···	6	7	8	9	10	11	12	13
Men										
Conditional	30	···	30	31	32	36	50	50	50	100
Prob. n occurs	100	···	100	99	94	84	61	33	33	23
Unconditional	30	···	30	30	30	30	30	16	16	23
Women										
Conditional	26	···	26	27	29	35	50	50	50	100
Prob. n occurs	100	···	100	98	90	75	52	26	26	15
Unconditional	26	···	26	26	26	26	26	13	13	15

Table 4.5: *Importance of game number in set*

Table 4.5 gives the values for each game number in a set, still based on equal probabilities of winning a service game (81.3% for men and 67.0% for women). It gives the *conditional* importance, $imp_{gs}(n)$; the probability of reaching the nth game, $\ell(n)$; and the *unconditional* importance that we need for the current hypothesis, $imp_{gs}(n) \times \ell(n)$.

It is now clear that there is nothing special about the seventh game. The conditional importance slowly increases with the game number from 30% (26% for the women) in the first game to 100% in the tiebreak. But if we take into account that the higher game numbers occur less frequently (for the men, 23% of the sets go to a tiebreak; 15% for the women), then there is still no special role for the seventh game. The three highest game numbers 11–13 are the most important *if they occur*, but since they do not occur that often, their unconditional importance is relatively small.

Big sets

To complete the three-step analysis (point-game, game-set, set-match), we consider the importance of a set in a match. We still assume that both players $\mathcal{I}$ and $\mathcal{J}$ have equal strength, so for each player the probability s_i to win the set is 50%. The importance imp_{sm} is given in Table 4.6.

		Men			Women	
		Set score j			Set score j	
		0	1	2	0	1
	0	38	38	25	50	50
Set score i	1	38	50	50	50	100
	2	25	50	100	—	—

Table 4.6: *Importance of set in match,* $s_i = 50\%$

In a best-of-three-set match, the importance is either 50% (at 0-0, 1-0, or 0-1) or 100% (at 1-1). The same applies to a best-of-five-set match if the first two sets have been shared, because at 1-1 the match becomes the same as a best-of-three match at 0-0.

At this point we need some new notation. We have used the notation $g_i(a, b)$ to denote the probability that $\mathcal{I}$ wins the game from the point score a-b, assuming implicitly that $\mathcal{I}$ is serving. Similarly, we now let $m_i(a, b)$ denote the probability that $\mathcal{I}$ wins the match at the set score a-b (with point and game scores both at 0-0). Whether $\mathcal{I}$ or $\mathcal{J}$ serves first in the match is not relevant.

What we earlier called m_i at the beginning of the match is thus the same as $m_i(0,0)$. (A list of all symbols used in the book is provided in Appendix B.)

The new elements in a best-of-five match concern 0-0, 1-0, 2-0, 0-1, and 0-2. Let us examine the importance of the third set at 2-0 in sets. If $\mathcal{I}$ wins the set, he automatically wins the match, formalized by $m_i(3,0) = 100\%$. If $\mathcal{I}$ loses the set, the score becomes 2-1, and he can win the match by winning the fourth set (with probability 50%) or by losing the fourth and winning the fifth set (with probability $0.5 \times 0.5 = 25\%$). Hence, $m_i(2,1) = 75\%$. The importance is

$$imp_{sm}(2,0) = m_i(3,0) - m_i(2,1) = 100\% - 75\% = 25\%.$$

The importance at 0-2 is the same, because

$$imp_{sm}(0,2) = m_i(1,2) - m_i(0,3) = 25\% - 0\% = 25\%.$$

The equality $imp_{sm}(2,0) = imp_{sm}(0,2)$ no longer holds when the players differ in strength. If $\mathcal{I}$ is stronger than $\mathcal{J}$, so that $s_i > 0.5$, then $m_i(2,1) > 75\%$ and $m_i(1,2) > 25\%$, implying

$$imp_{sm}(2,0) < imp_{sm}(0,2).$$

The third set is thus more important (for both players!) if the weaker player has won the first two sets than if the stronger player has won them. Similarly, the second set is more important if the weaker player leads than if the stronger player leads.

Are all points equally important?

One sometimes hears that

Hypothesis 5: *All points are equally important.*

This is not true. Consider, for example, the famous match between John Isner and Nicolas Mahut, analyzed on pages 24–27. Isner won the match but scored twenty-four points fewer than Mahut. This can only happen if points are *not* equally important. Importance varies across points. Big points do exist.

We can say more by using the concept of importance. In a game, breakpoint is a more important point than 40-0. Games at the end of a set are more important than at the beginning. The final set is more important than the first set.

We have defined the importance of a point in a game (imp_{pg}), of a game in a set (imp_{gs}), and of a set in a match (imp_{sm}). These definitions and the rules of conditional probability now imply that the importance of a point in a match (imp_{pm}, also written simply as imp) can be expressed as

$$imp = imp_{pm} = imp_{pg} \times imp_{gs} \times imp_{sm}.$$

This is convenient, because the complicated importance of a point in a match can thus be calculated as the product of three much less complicated importance measures. We have seen that there is substantial variation in the values of these importance measures, and hence there is also substantial variation in the importance of a point in a match. This confirms our rejection of hypothesis 5.

The most important point

What then is the most important point in a match? Matchpoint perhaps? Usually not. For example, matchpoint at 2-0, 5-0, 40-0 is not important at all, because even if player $\mathcal{I}$ loses that point, he will almost certainly win the match. On the other hand if we are in the tiebreak of the final set (2-2 for the men and 1-1 for the women) with a score of 6-5 or 5-6, then matchpoint will be an important point.

The most important point is the point where the swing between winning the point and losing it is largest. Points at the end of a tiebreak have this property. Also breakpoints are important, as Tables 4.1 and 4.2 show. So we expect high importance values for breakpoints, particularly in the final set.

Table 4.7 confirms this qualitative analysis. It reports the importance imp of various points in a match when $\mathcal{I}$ is serving, again taking $p_i = p_j = 64\%$ for the men and $p_i = p_j = 57\%$ for the women. The final set has a tiebreak, as at the US Open. Breakpoint in the final set at 5-4, 30-40 has an importance of 38% for the men ($0.76 \times 0.5 \times 1 = 0.38$). This is about three times as important as matchpoint at 5-4, 40-30 ($imp = 12\%$), and it is also more important than the points in the fourth-set tiebreak. Points in a final-set tiebreak and breakpoints towards the end of the final set

Game score	Point score	Men Set score 2-2	Men Set score 2-1	Women Set score 1-1	Women Set score 1-0
6-6	6-5	50	25	50	25
6-6	5-6	50	25	50	25
5-4	40-30	12	6	18	9
5-4	30-40	38	19	32	16
4-5	40-30	12	6	18	9
4-5	30-40	38	19	32	16

Table 4.7: *Importance of point in match,* $p_i = p_j =$ 64% *(men) and* 57% *(women)*

are therefore the most important points in a match. Comparing men and women in the table, we see that the ordering of the *imp* values is the same, but the values themselves are not, because the point probability p is higher for men than for women.

Do men play more or fewer important points than women in a match? To answer this question we need to realize that the results in Table 4.7 are conditional on the occurrence of an important point. Women will play more final sets than men, because the women play best-of-three sets and the men best-of-five. But the men encounter important points more frequently if they play a final set, because break chances occur less frequently, so that if they occur they are more important. Similar to the result concerning long matches on page 25, these two effects roughly offset each other: men and women play about the same number of important points.

Three importance profiles

We now consider the full sequence of points for the same three matches as in the previous chapter, this time not from the viewpoint of forecasting but from the viewpoint of importance. In each of the three matches the biggest (most important) point occurs at breakpoint towards the end of the final set. The biggest point of all is breakpoint in the fifth set of the 2008 Wimbledon final between Roger Federer and Rafael Nadal with an importance of 42.2%.

Federer-Nadal

In the 2008 Wimbledon final, Roger Federer was defeated by Rafael Nadal in five sets: 4-6, 4-6, 7-6, 7-6, 7-9. A total of 413 points were played. Figure 4.1 confirms the large variation in importance across points. Eighty of the points had *imp* > 10 and four really big points had *imp* > 40, all in percentages.

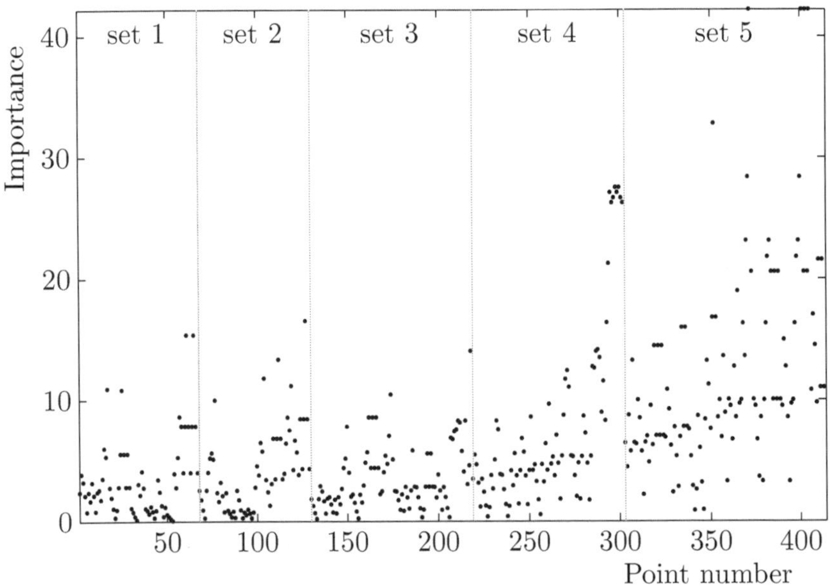

Figure 4.1: *Importance profile, Federer-Nadal, Wimbledon 2008*

In the first set, there were breakpoints at 1-1 and 1-2, and two further breakpoints at 4-5. At 4-5, Nadal served for the set and Federer had two breakpoints to level at 5-5 (*imp* = 15.4). Nadal's setpoints (*imp* = 4.0) were less important than Federer's breakpoints. In the second set there were breakpoints at 4-2, 4-3, 4-4, and 4-5, all visible in the figure. In both sets, Nadal was serving for the set at 4-5, and in both cases Federer had breakpoints. In the first set, the importance of the breakpoint was 15.4; in the second set 16.5. This is because the second set is more important than the first when the underdog (Nadal in this case) wins the first set; see the discussion on page 57.

The only big point in the third set occurred at 6-5 in the tiebreak with Federer serving for the set (imp = 14.0). Many big points occurred in the fourth-set tiebreak: at 5-5 and 7-7 with Federer serving (imp = 27.1) and at 6-6 and 8-8 with Nadal serving (imp = 26.7). Nadal's points are slightly less important than Federer's because if Nadal loses his service point it is likely that he loses the set but not necessarily the match, while if Federer loses his service he will likely lose the match. There are two setpoints for Federer (imp = 26.3) and two matchpoints for Nadal (imp = 27.5).

In the fifth set the importances are higher overall than in the fourth, corresponding to the larger shocks in the forecasting profile in the fifth set compared to the fourth (Figure 3.1). The breakpoint at 4-3 on Nadal's service has imp = 32.8. But the four really big points all occur later in the fifth set, all on Federer's service: the breakpoint at 5-5 and three further breakpoints at 7-7 (imp = 42.2). Having broken at 7-7, Nadal serves for the match. Matchpoint (imp = 11.0) is rather less important than many of the previous points.

Clijsters-Williams

The semi-final at the 2010 US Open between Kim Clijsters and Venus Williams was won by Clijsters in three sets: 4-6, 7-6, 6-4. Of the 191 points played, there were 32 points with $imp > 10$, 3 of which with $imp > 20$. In the first set, the biggest point was breakpoint at 3-3 on Clijsters' service (imp = 11.9), which Williams won, going on to win the set. This is visible in Figure 4.2 as the highest dot in the first set. The second-set tiebreak produced the big points in the second set, interestingly at the beginning (0-0 and 1-0). The tiebreak went to 4-0 and 7-2, not producing any other big points.

In total there were nine points with $imp > 15$, all in the third set. Early in the third set there were breakpoints on Clijsters' service at 0-1 (imp = 17.3) and at Williams' service at 1-1 (imp = 15.8). The three biggest points occurred at 4-3 when Williams had a breakpoint (imp = 22.2), and at 4-4, Williams serving, at 30-30 (imp = 23.6) and at breakpoint (imp = 28.0). At 5-4 Clijsters won her first matchpoint at 40-15, but this was not a big point (imp = 6.8).

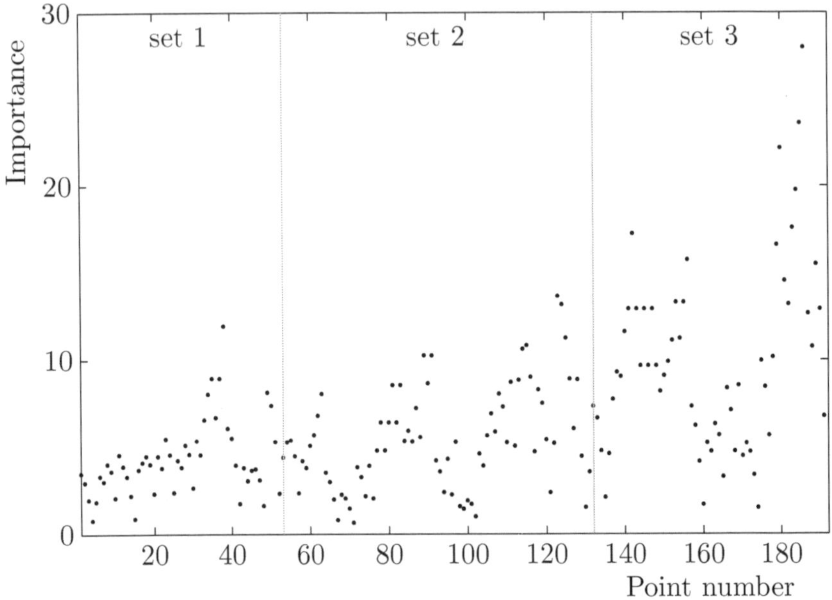

Figure 4.2: *Importance profile, Clijsters-Williams, US Open 2010*

Djokovic-Nadal

Finally we reconsider the 2012 final at the Australian Open, where Novak Djokovic defeated Rafael Nadal in five sets: 5-7, 6-4, 6-2, 6-7, 7-5. The importance profile, shown in Figure 4.3, provides the importance of each of the 369 points played. Fifty-six points had an importance of $imp > 10$, sixteen of $imp > 20$, and four of $imp > 30$. Two big points occurred in the fourth-set tiebreak, and all other big points occurred in the fifth set. There were breakpoints at 2-3 ($imp = 27.6$) and at 2-4 ($imp = 26.2$). The biggest points occurred at 4-4 and 5-5, Nadal serving, breakpoint to Djokovic ($imp = 34.7$), and at 6-5, Djokovic serving, breakpoint to Nadal ($imp = 34.6$).

Further reading

The seminal paper on importance in tennis is Morris (1977), and our definition and several other issues in the current chapter build on this paper. Morris also notes that the aggregate importance of

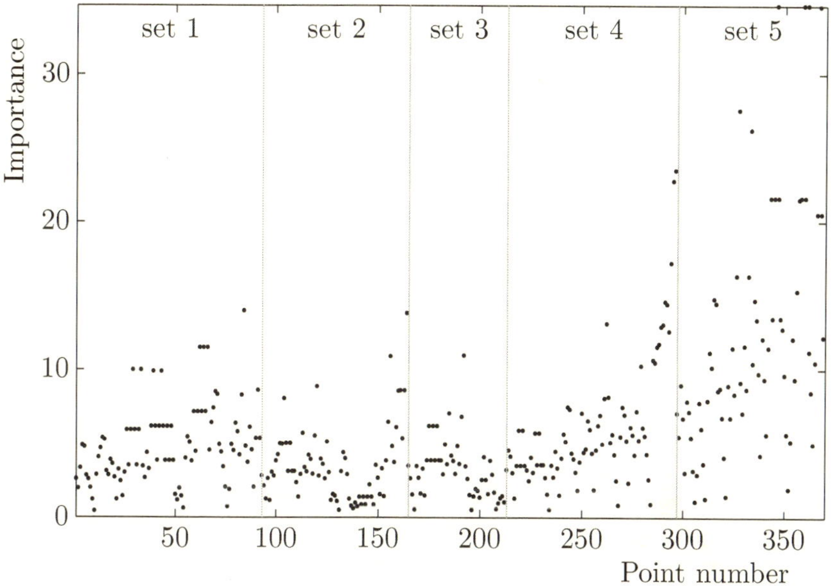

Figure 4.3: *Importance profile, Djokovic-Nadal, Australian Open 2012*

all points served to the deuce court equals that for the ad court. Because more points are served to the deuce court, higher *average* importance is experienced in the ad court. This is one reason for double teams to have the player who performs best at important points receive on the left side.

Another notable result is that the player who serves first in a set serves in the less important games. This does not imply that serving first is an advantage in the sense that the server wins his or her games more easily. In Chapter 11 we will test whether such an advantage exists. If it does exist, then it could possibly be explained by the reduced importance and pressure.

5 Point data

So far we have used very few data and no statistical techniques; we have relied on mathematics. If we want to prove or disprove popular tennis hypotheses (e.g., 'real champions win the big points'), find out more about the trade-off between first and second service, or determine whether 'momentum' or 'winning mood' exist in tennis, then we need to apply statistical techniques. But before we can apply these techniques we need data.

We shall use various data sets, some recent, some less recent. In this chapter we describe a Wimbledon data set at point level over four years, 1992–1995. This will be our main data source. It is a rich source covering many different players. But it is not the individual players that we are primarily interested in. Our interest is in professional tennis in general, and the Wimbledon data set will teach us many lessons about professional tennis. An obvious concern is whether these relatively old data are still of interest today. We shall study and comment on the changes over time, using other data sets, and conclude that our Wimbledon data set is still remarkably relevant.

The Wimbledon data set

Our data set consists of 481 matches (88,883 points) played in the men's singles and women's singles at Wimbledon from 1992 to 1995. This accounts for almost half of all singles matches played during the four years, as only matches on one of the five 'show courts' were recorded (the show courts are the courts where the matches with the best-known players are typically scheduled). For each of these matches we know the players, their rankings, and the exact sequence of points. We also know at each point whether or not a second service was played and whether the point was decided through an ace or a double fault.

	Men	Women
Matches	258	223
Sets	950	503
Final sets	51	57
Games (excl. tiebreaks)	9367	4486
Tiebreaks	177	37
Points (incl. those in tiebreaks)	59,466	29,417
Points (excl. those in tiebreaks)	57,319	28,979

Table 5.1: *Wimbledon data set: totals, 1992–1995*

Table 5.1 provides a summary of the data. We have slightly more matches for men (258) than for women (223), but of course many more sets, games, and points in the men's singles than in the women's singles, because men play best-of-five sets and women best-of-three. Almost 60,000 points are observed in the men's singles and almost 30,000 in the women's singles. The men's singles are therefore seriously overrepresented on the show courts, and about two-thirds of the spectators' viewing time goes to the men. If sex equality in viewing time would be a goal, The All England Lawn Tennis and Croquet Club should aim for 193 matches (a 25% reduction) in the men's singles and 335 matches (a 50% increase) in the women's singles on the five show courts. (In Wimbledon's defense: a slightly smaller adjustment will achieve equality in viewing time, due to the fact that women play longer rallies than men.)

Table 5.2 presents proportions associated with the totals in Table 5.1. The men play on average 231 points per match, the women 132, and hence a match in the men's singles takes on average 1.75 times as long (in terms of points) as a match in the women's singles.

Both men and women play about the same number of points per set, around sixty. Because the men have a higher probability of winning a service point than the women, the men play fewer points per game than the women, but they play more games per

	Men	Women
Sets in match	3.7	2.3
Games in set	9.9	8.9
Games in match	36.3	20.1
Tiebreaks in (non-final) set	0.2	0.1
Tiebreaks in match	0.7	0.2
Points in game	6.1	6.5
Points in tiebreak	12.1	11.8
Points in set	62.6	58.5
Points in match	230.5	131.9

Table 5.2: *Wimbledon data set: proportions, 1992–1995*

set. These two opposite forces appear to be approximately equal, resulting in approximately the same number of points per set. In addition, the quality difference in the men's singles is smaller than in the women's singles, as discussed on page 24: scores like 6-0 and 6-1 are more common for the women than for the men. This is why there are more tiebreaks per set for men than for women.

The original data set was of high quality with very few errors. Still, after performing all possible checks, we had to delete about 4% of the matches, which contained non-repairable errors or had not been completed. The remaining 481 matches constitute our data set — they passed all our tests and are therefore 'error-free'. For a statistician, having a large, detailed and error-free data set is an exceptional and happy situation. In most studies the reliability of the data is a serious concern, but in sport statistics the data are typically of very high quality, allowing better analysis and sharper conclusions.

Two selection problems

All matches in our data set are played on one of the five show courts: Centre Court and courts 1, 2, 13, and 14 at the time. Because top players are typically scheduled on these courts, this causes an underrepresentation in the data set of matches involving non-seeded players. This is a problem, a so-called selection problem.

It is not the only selection problem. If two non-seeded players meet in the quarter-final, then this match is likely to be scheduled on a show court. But, if they play each other in the first round, their match is considered to be less important and is likely to be played on another court. After all, there are sixteen first-round matches involving a seeded player and such matches typically take precedence. (In the period 1992–1995, sixteen players were seeded; from 2001 onwards, thirty-two.) The underrepresentation of matches between two non-seeded players is therefore most serious in the early rounds. This dependence on round in the selection of matches is also present in other types of matches (seeded versus non-seeded or seeded versus seeded), but there the problem is less serious.

	Sd-Sd		Sd-NSd		NSd-NSd		Total	
Round	S	P	S	P	S	P	S	P
Men								
1	—	—	48	64	34	192	82	256
2	—	—	46	54	16	74	62	128
3	—	—	39	41	16	23	55	64
4 (last 16)	8	9	15	15	8	8	31	32
5 (quarter)	7	7	9	9	0	0	16	16
6 (semi)	7	7	1	1	0	0	8	8
7 (final)	4	4	0	0	0	0	4	4
Total	26	27	158	184	74	297	258	508
Women								
1	—	—	43	63	24	193	67	256
2	—	—	43	58	3	70	46	128
3	—	—	42	48	12	16	54	64
4 (last 16)	8	8	20	21	2	3	30	32
5 (quarter)	11	12	3	3	1	1	15	16
6 (semi)	6	6	1	2	0	0	7	8
7 (final)	4	4	0	0	0	0	4	4
Total	29	30	152	195	42	283	223	508

Table 5.3: *Matches between seeded (Sd) and non-seeded (NSd) players in the sample (S) and in the population (P)*

Table 5.3 provides more details about both selection problems. We distinguish between round (1 = first round, ..., 7 = final) and type of match (Sd-Sd for two seeded players, Sd-NSd for a seeded against a non-seeded player, and NSd-NSd for two non-seeded players). The columns labeled 'S' contain the number of matches in our sample, and the columns labeled 'P' give the number of all matches played (matches in the population). In the first round of the women's singles there are sixty-three rather than sixty-four matches between a seeded and a non-seeded player. This is because Mary Pierce, seeded 13, withdrew in 1993 at the last moment. She was replaced by Louise Field, an unseeded player.

The percentage of matches involving two non-seeded players (NSd-NSd) in our data set is 24.9 (74/297) for the men and 14.8 (42/283) for the women. Both are lower than the percentages for Sd-NSd matches, which are themselves lower than those for Sd-Sd matches. This illustrates the first selection problem: the underrepresentation of matches involving non-seeded players.

Round dependence, the second selection problem, is caused by the increasing sampling percentages over the rounds. For example, only 32.0% (82/256) of all first-round matches in the men's singles and 26.2% (67/256) in the women's singles are in the data set, whereas all finals have been sampled.

Since we wish to make statements about Wimbledon in general, and not just about the matches in our sample, we account for both selection problems by weighting the matches when computing statistics. The weight of a match is the ratio P/S using the population and sample values in Table 5.3 for the round and type of match. The validity of this procedure involves an assumption, namely that the decision by Wimbledon's organizers whether a match in one cell is on a show court or not is random within that cell, so that the matches on the show courts (which are the matches we observe) are representative.

One could argue that, if the sample is very small compared to the population, this method would make the few observed matches too important. Most notably, in the women's singles we observe only three of the seventy matches played between two non-seeded players in the second round. If these three matches were selected by the organizers to include, for example, players just outside the top sixteen, then our method would be seriously biased for this cell. As it happens, the three matches concern players with WTA rankings 27-41, 131-143, and 22-113, and hence there is no reason to believe that these matches are not representative. Still, in the more advanced analyses from the next chapter onwards we combine the three women's matches in the second round with the twenty-four matches in the first round, and weight all twenty-seven matches by 263/27.

We have experimented with other weighting methods, and we find that our results are not sensitive to the method of weighting. Whenever the Wimbledon data set is used in this book, we shall use the above weights to make the sample more representative and we shall call the weighted data set simply 'the data set'.

Estimators, estimates, and accuracy

One concern with our detailed data set is that it is relatively old. We consider the years 1992–1995 and one may wonder whether these data are still of interest today. Has tennis not changed? This is a reasonable concern and we need to address it.

We start by considering p, the probability of winning a service point. This was the key probability in previous chapters, where we allowed it to differ across players, reflected in the notation p_i. In the following chapters we will again differentiate between players, for example by introducing a player's 'quality' in Chapter 7. But in the current chapter we simply put all players together. Hence there is one single p for the men and one single p for the women, reflecting the service strength of professional tennis players in general.

The value of p is not observable. To estimate it we need data and a bit of statistical theory. Suppose there are T service points in the sample. Some of these T points will be won by the server (successes), the other points by the receiver. A natural estimator of the unknown probability of success p is the observed relative frequency of successes:

$$\hat{p} = \frac{\text{number of service points won by server}}{\text{total number of service points in the sample}},$$

where the hat above p signifies that we are dealing with an estimator. We have $T = 59,466$ for the men and $T = 29,417$ for the women, and we obtain $\hat{p} = 64.4\%$ and $\hat{p} = 56.1\%$, respectively. Not surprisingly, men win more points on service than women.

To be precise: an estimator is a formula, like the formula for $\hat{p}$ above. An estimate is the realization of the formula when we substitute numbers for symbols. So, the realizations $\hat{p} = 64.4\%$ and $\hat{p} = 56.1\%$ are estimates, not estimators.

This estimator of p is a good estimator in the sense that on average it produces the right answer. We express this by saying that the expected value of $\hat{p}$ is equal to the true value: $\mathrm{E}(\hat{p}) = p$. But this is only on average. There will be deviations from the average, $\hat{p} - \mathrm{E}(\hat{p})$, and the variation of $\hat{p}$ is usually expressed as the variance,

$$\mathrm{var}(\hat{p}) = \mathrm{E}\left(\hat{p} - \mathrm{E}(\hat{p})\right)^2.$$

To compute the variance, we put all players together (only in this chapter). This means that we assume that *all* points are iid, not only within players (as in hypothesis 1), but also between players. It is doubtful whether the latter part of the assumption is realistic, and we shall return to it in the next chapter. Assuming that both parts of the assumption are satisfied, the variance can be computed as $\mathrm{var}(\hat{p}) = p(1-p)/T$. The higher is T, the lower is $\mathrm{var}(\hat{p})$, the closer is $\hat{p}$ to the true value p, and the more accurate is $\hat{p}$.

Since the variance involves the square of p, we typically take the square root of the variance in order to obtain a measure in the same scale as p. The estimate of this square root is called the standard error, written as 'se'. To quantify the inaccuracy of the estimate we thus take

$$\mathrm{se} = \sqrt{\hat{p}(1-\hat{p})/T},$$

where we have substituted the unknown p by its estimate. In our case the standard errors are se $= 0.2\%$ (men) and se $= 0.3\%$ (women), smaller for the men than for the women, mainly because we observe more points for the men.

The estimation results are then summarized as

$$\hat{p} = \begin{cases} 64.4\%\ (0.2\%) & \text{for the men,} \\ 56.1\%\ (0.3\%) & \text{for the women.} \end{cases}$$

Under certain conditions, which are typically assumed to hold, the interval defined by the estimate $\pm$ twice the standard error covers 95% of the possible outcomes of the estimator. This means, loosely speaking, that we expect p to lie between 64.0% and 64.8% for the men and between 55.5% and 56.7% for the women. These uncertainty intervals are called 'confidence intervals'. The confidence intervals are quite narrow here, reflecting the precision of our estimates, caused by the fact that we have many observations. Statistical theory has thus helped us to learn about the magnitude of something unobservable, the probability of service success, as well as about the accuracy of that information.

Development of tennis over time

To answer the question of whether the point data are still of interest today, we study the two relative frequencies 64.4% (men) and 56.1% (women) in some more detail. How stable are these estimates over time?

	Men		Women	
1992	64.9	(0.4)	57.0	(0.6)
1993	64.9	(0.4)	56.6	(0.6)
1994	63.9	(0.4)	55.4	(0.6)
1995	64.0	(0.4)	55.4	(0.5)

Table 5.4: *Percentage of points won on service, 1992–1995*

Table 5.4 presents the development of p during our observation period. The numbers in brackets denote again the standard errors of the estimators. The estimates suggest that service dominance has decreased somewhat between 1992 and 1995, both for men and for women, presumably not because the service has weakened but because the return of service has improved. However, a value of, say, 64.5% lies within all four uncertainty intervals for the men, so that the true p may in fact have been constant over the years. In that case, the decrease in the service dominance may be spurious and could just be a coincidence. The same holds for, say, 56.0% in the women's singles.

Has this (possible) decrease of the service dominance continued? Using totals over all Wimbledon matches from 1992 until 2010 (in-

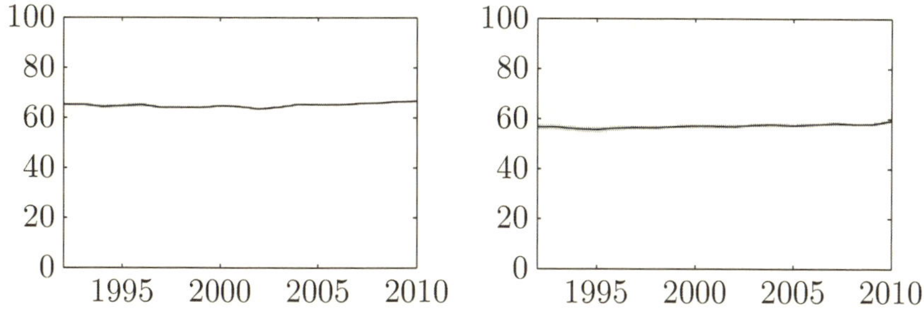

Figure 5.1: *Service dominance, 1992–2010 (men left, women right)*

formation that is not in our main data set), we plot the estimated value of p in Figure 5.1. There is no indication of a further decrease in service dominance. The probability of winning a service point appears to be remarkably stable over the years.

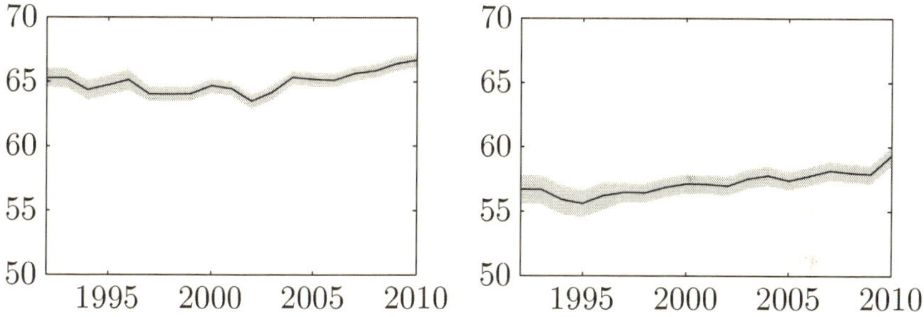

Figure 5.2: *Service dominance, 1992–2010, zoomed in (men left, women right)*

Zooming in on Figure 5.1 produces Figure 5.2. Here some more detail is visible, including the 95% uncertainty bounds. The probabilities are still quite stable, and there is no indication of a further decrease of service dominance. If anything, it appears that the service dominance is increasing from 2003 onwards for the men, and from 1995 onwards for the women, but the evidence is not strong.

The 1992–1995 data are therefore quite informative about tennis in later years, at least regarding the probability p of winning a point on service. This does not mean that tennis has been stable in every dimension. For example, there has been a decline of serve-and-volley tennis — the grass remains greener near the net during

Wimbledon — but this has not coincided with a drop in service dominance p. Maybe the service has become more powerful. If so, then the return of service has also improved, and both effects apparently offset each other.

Winning a point on service unraveled

The service is one of the most important aspects of tennis, particularly on fast surfaces such as the grass courts at Wimbledon. So far we have only discussed the probability p of winning a point on service. But tennis has two serves, and therefore more information can be obtained from the data.

	Men		Women	
1st service in	59.4	(0.2)	60.8	(0.3)
2nd service in	86.4	(0.2)	86.0	(0.3)
Points won if 1st service in	73.3	(0.2)	62.2	(0.4)
Points won if 2nd service in	59.4	(0.3)	54.1	(0.5)
Points won on 1st service	43.6	(0.2)	37.8	(0.3)
Points won on 2nd service	51.4	(0.3)	46.6	(0.5)
Points won on service	64.4	(0.2)	56.1	(0.3)

Table 5.5: *Service percentages*

The principal service characteristics are provided in Table 5.5. Some of these characteristics are typically also shown on television, although they are sometimes called differently. For example, the commonly presented 'winning % on 1st serve' and '1st serve points won', two names for the same statistic, are what we call 'points won if 1st service in', which is a more accurate description. What we call 'points won on 1st service' is the percentage of 1st services hit that have resulted in winning the point. This corresponds to the earlier definition of 'points won on service' (the percentage of points served that have resulted in winning the point).

The characteristic that is typically *not* shown on television is 'points won on 1st (or 2nd) service'. This is remarkable, because it is precisely this probability that is of most interest. After all, almost anyone (even amateur players) can achieve a '1st (2nd) service in' percentage of close to 100%, although such a service would be too easy for the receiver and therefore not optimal. Most professional players can also achieve a high percentage on 'points won if 1st (2nd) service is in', by making the service very risky, so that it is a fault most of the time, but if it goes in, it will have a high chance of winning the point. Both probabilities are typically presented on television. But it is not the two probabilities themselves that matter most — it is their product. As a result, the percentage 'points won on 1st (or 2nd) service' provides a proper measure of the dominance of the first (second) service.

In the men's singles, on average, the first service is in 59.4% of the time. If the first service is in, the probability of winning the point is 73.3%. The probability of winning the point on the first service is therefore 59.4% × 73.3% = 43.6%. In general,

$$\begin{aligned}&\%\text{ of points won on 1st service}\\&\qquad = (\%\text{ of points won if 1st service in}) \times (\%\text{ 1st services in}).\end{aligned}$$

Of course, the same holds for the second service.

Combining the data for the first and second services, we can derive the percentage of points won on service, the correct measure of service dominance. A player can win a point on service in two ways: on the first or on the second service, where the second possibility only becomes relevant when the first serve is a fault. Therefore,

$$\begin{aligned}&\%\text{ of points won on service} = \%\text{ of points won on 1st service}\\&\quad + (\%\text{ 1st service not in}) \times (\%\text{ of points won on 2nd service}).\end{aligned}$$

For example, in the men's singles,

$$64.4\% = 43.6\% + (100 - 59.4)\% \times 51.4\%.$$

The percentage of points won on the first service is plotted in Figure 5.3, where we have 'zoomed in' to provide more detail. The probabilities are quite stable, although they have risen slightly in recent years, both for men and for women. The second-service percentages, presented in Figure 5.4, are even more stable over time than the first-service percentages.

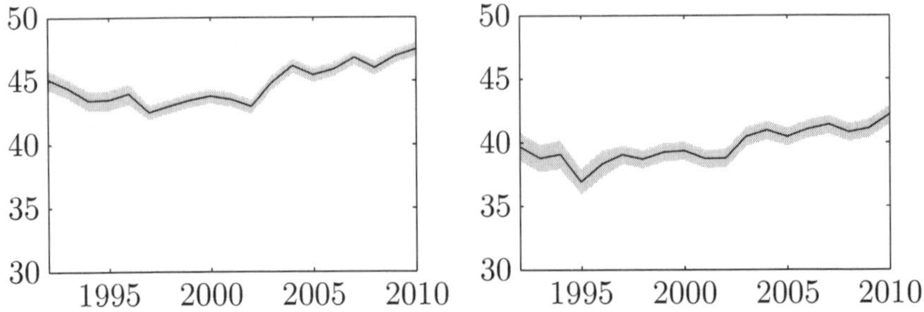

Figure 5.3: *Percentage of points won on first service, 1992–2010 (men left, women right)*

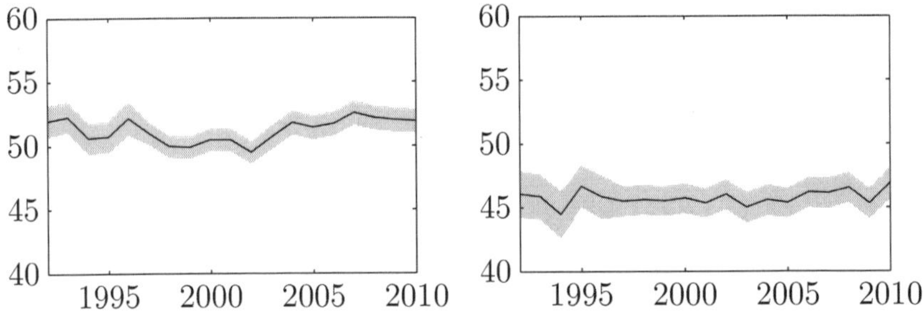

Figure 5.4: *Percentage of points won on second service, 1992–2010 (men left, women right)*

Testing a hypothesis: men versus women

A comparison between men and women in Table 5.5 reveals — apart from the obvious differences — that the percentage of first and second service in is almost the same. The first service is in 59.4% of the time for men and 60.8% for women, while the second service is in 86.4% of the time for men and 86.0% for women. It is not immediately clear why these percentages should be so close.

What do we actually mean by 'almost the same'? To answer this question we formulate the following hypothesis.

Hypothesis 6: *The probability that the service is in is the same in the men's singles as in the women's singles.*

To test the hypothesis we consider the difference diff_1 between men and women of the probability that the first service is in. Similarly, we consider the difference $diff_2$ for the second service. If the hypothesis is true, then $diff_1$ and $diff_2$ are both zero. We don't know $diff_1$ and $diff_2$, but from Table 5.5 we can derive estimates of them, namely −1.40 and 0.38, respectively.

The fact that these numbers differ from zero does not necessarily mean that we reject the hypothesis. It could be pure randomness that has made them different from zero even though the hypothesis is actually true. The question is whether they are *sufficiently* different from zero to reject the hypothesis. To understand what 'sufficiently' means in a statistical sense, we must take the uncertainty of the estimates into account, and this is where we need the standard error of the estimators. Because the men and women samples are independent (an assumption made throughout this book), the variance of the estimator of $diff_1$ (and similarly of $diff_2$) is equal to the sum of the variance of the estimator for the men and the variance of the estimator for the women. We then find from Table 5.5 that

$$\widehat{diff}_1 = -1.40\ (0.34), \qquad \widehat{diff}_2 = 0.38\ (0.40).$$

Given these estimates and standard errors we can construct confidence intervals similar to those on page 72. In this case, the confidence intervals are (−2.08, −0.72) and (−0.42, 1.18), respectively. This means that with 95% certainty the data imply that the true $diff_1$ and $diff_2$ lie inside these intervals.

If the hypothesis were true, then we would expect that zero lies inside the interval. This is the case for the second service, and hence there is no reason to reject the hypothesis that $diff_2 = 0$. If, on the other hand, zero lies outside the interval (as is the case for the first service), then something is wrong. Many things could be wrong, but if we assume that all underlying assumptions hold except possibly the hypothesis, then we must conclude that the hypothesis is wrong and reject it.

Given the 95% confidence, we will incorrectly reject the hypothesis 5% of the time. So in 5% of the cases we reject a hypothesis when in fact it is true. This is the price one has to pay in hypothesis testing. All testing in the book is based on the 5% level.

Another way of saying that we reject the hypothesis, is to say that the underlying estimate, here $\widehat{diff}_1$, is 'statistically significantly' different from zero, in short 'significant'. If, on the other hand, zero lies inside the interval (as for the second service), then there is no evidence that something is wrong and we do not reject the hypothesis. The underlying estimate, here $\widehat{diff}_2$, is then called 'insignificant'.

Insignificance does not mean that we accept the hypothesis. Statisticians always emphasize that hypothesis testing is about rejecting or not rejecting a hypothesis, not about rejecting or accepting a hypothesis. This is because, if we do not reject a hypothesis, there are many possible reasons and only one reason is that the hypothesis is true. It could be, for example, that the statistical analysis is not powerful enough to detect deviations from the hypothesis even when the hypothesis is in fact false.

Hypothesis testing results in a binary outcome: rejection or no rejection. This is not completely satisfactory, because a small increase in standard error could change the conclusion from reject to not reject. The binary nature uses only part of the information available in the data, in contrast to confidence intervals. Still, a binary outcome is customary and simple, and we will use it throughout this book.

This is hypothesis testing in a nutshell. Based on our estimates and assumptions we conclude that the probability that the first service is in is not the same in the men's singles as in the women's singles, but that the probability that the second service is in may be the same.

Aces and double faults

Of special interest, and always included in television statistics, are the number of aces and double faults. The percentage of aces is defined as the ratio of the number of aces (first or second service) to the number of points served, rather than to the number of services. The percentage of double faults is the ratio of the number of points with a double fault to the number of service points.

In our Wimbledon data set, 8.2% (0.1%) of the points were decided through an ace in the men's singles, and 3.1% (0.1%) in the women's singles. It is not surprising that men serve almost three times as many aces as women. More surprising perhaps is the stability of this percentage over time, as shown in Figure 5.5.

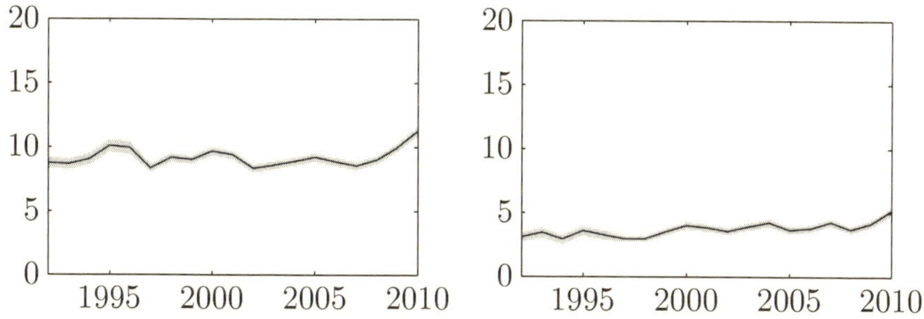

Figure 5.5: *Percentage of aces, 1990–2010 (men left, women right)*

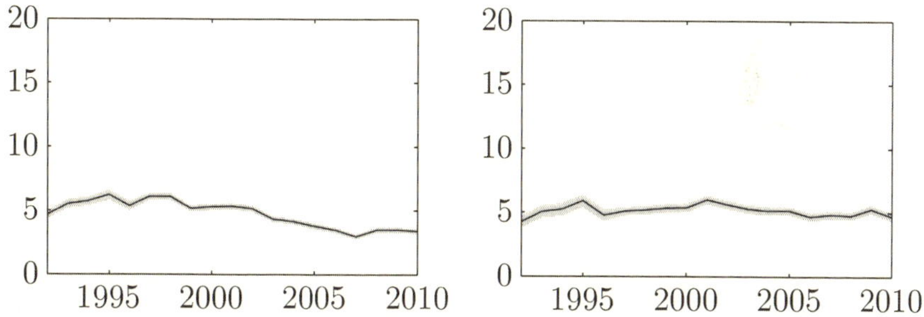

Figure 5.6: *Percentage of double faults, 1990–2010 (men left, women right)*

Figure 5.6 shows that the percentage of double faults has decreased slightly over time for the men, which could still be due to estimation uncertainty, but has remained stable for the women. About 5.5% of the points served result in a double fault, both for the men and for the women. This can be deduced from Table 5.5, because

$$\begin{cases} (1-0.594)\times(1-0.864) = 5.52\% & \text{for the men,} \\ (1-0.608)\times(1-0.860) = 5.49\% & \text{for the women.} \end{cases}$$

The standard errors are 0.09% and 0.13%, respectively.

An alternative to hypothesis 6, now taking both services into account, is to compare double faults.

Hypothesis 7: *The probability of a double fault is the same in the men's singles as in the women's singles.*

The difference in probability between men and women is estimated to be 0.03%-points with a standard error of 0.16%-points. Since the standard error is large relative to the estimate, the confidence interval (−0.29, 0.35) covers zero, so that the hypothesis is not rejected. Therefore, it may indeed be the case that the percentage of double faults is the same for men and women. This is a remarkable fact, which requires further investigation. Unfortunately, our data are not sufficiently rich to understand the underlying cause.

Of course, players could lower the number of double faults by making their service easier. They don't do this, because the probability of winning a point decreases if the service is too easy. If a player serves *no* double faults in a match, this does not necessarily show that he or she has served well. It may well indicate that the server has not taken enough risk at his or her service. There is an optimal number of double faults in each match for each player, and this optimum number is, in general, not zero.

Breaks and rebreaks

So far we have only used simple frequencies, and this has produced credible estimates and test outcomes. But this need not always be the case. Unexpected conclusions can emerge if we do not analyze the data in a statistically sound manner. Consider for example the famous break-rebreak hypothesis.

A break occurs if the server does not win his or her own service game. If this break does not decide the set, then the next game is often considered special. The server in the next game can now 'confirm' the break by holding service, or he or she can be broken back. Some commentators believe that there is a higher probability of being broken back after a break, perhaps because the player who achieved the break enjoys the success and relaxes a bit, while the opponent is eager to strike back immediately.

Hypothesis 8: *After a break the probability of being broken back increases.*

If the hypothesis is true, then the probability of winning a point on service decreases after a break in the previous game (in the same set). Using our Wimbledon data we obtain Table 5.6. In the men's singles, 64.4% of the points are won by the server. The women win fewer service points (56.1%), as we already know. If the previous game was a break, then the probability of holding service is larger rather than smaller, both for men and women. And, if the previous game was not a break, then the probability is smaller than average. The difference between 'after break' and 'after no-break' is 1.7%-points (men) and 3.0%-points (women), respectively. Both estimates are significant.

	Men		Women	
All points	64.4	(0.2)	56.1	(0.3)
After break	65.8	(0.5)	57.9	(0.5)
After no-break	64.1	(0.2)	54.9	(0.4)

Table 5.6: *Percentage of winning a point on service after winning or losing the previous game*

If we translate these percentages from points to games, then we find that the probability of winning a service game is 3.3%-points higher if a break occurred in the previous game than if no break occurred. This is for the men; for the women the difference is 5.7%-points.

It seems therefore that the hypothesis is wrong. The opposite is true. But is this a credible conclusion? For example, should we not allow for quality differences between players? If the top seed plays against a relatively weak player and wins 6-0, 6-0, 6-0, then there are many breaks but no rebreak, not because the hypothesis is wrong, but because the top seed is a much better player. The 'after break' winning probability is thus dominated by good players and therefore higher than the 'after no-break' winning probability, which is dominated by weaker players. Quality heterogeneity can explain a positive difference, even when a break in the previous game does not matter at all.

This is an example of a well-known issue in statistics called sample selection bias: players are not randomly selected into the sample used to estimate a probability and this biases the estimators. Sample selection bias matters for many simple statistical analyses, also in this book. Sometimes the impact is small, sometimes it is big. The problem is that we do not know this beforehand. We shall return to the break-rebreak hypothesis and investigate the effect of sample selection bias in Chapter 12, after we have developed a statistical model that properly handles quality differences to avoid the bias.

Are our summary statistics too simple?

One of the purposes of this little book is to show that sometimes a simple statistical analysis suffices, and sometimes not. In the break-rebreak case, the analysis is 'too simple'. To find the precise balance between complexity and simplicity is the art of modeling.

This chapter has introduced our data set and we have performed some simple averaging exercises. Two things, in particular, were ignored: clustering and quality. We have treated every point the same, while in fact the points are clustered in matches and, within a match, in players. Each player has his or her own underlying probabilities and the performance of one player depends on the opponent in the match. Moreover, the differences between players are not random. Some have a higher ranking than others and we want to exploit such observable information in order to obtain better estimates. How do we allow for these quality differences? And, more difficult, how do we allow for unobservable quality differences, such as form of the day? Chapter 6 will deal with the clustering issue; the quality correction is discussed in Chapter 7.

Further reading

Summary statistics are often presented on television. Bedford *et al.* (2010) discuss many variants and their interpretations. They also show how summary statistics vary across surfaces. For example, both men and women win the highest percentage of service points when playing on grass, followed by carpet, hard court, and clay.

O'Donoghue and Ingram (2001) examine other types of summary statistics. They analyze more than three hundred hours of grand slam matches over the period 1997–1999, regarding the number of shots per rally, rally time, inter-point time, and so on. Rallies in men's matches are shorter than in women's matches, they are shortest at Wimbledon, and longest at Roland Garros. Men take more time between points than women. We do not know whether this still holds today.

Several authors study the development and changes of tennis over time. Coe (2000) relates the changes in tennis over many decades to technology, from rackets and balls to court surfaces. Guillaume *et al.* (2011) examine the careers of the top-ten men and women players between 1973 and 2009. The average career length is about sixteen years for both men and women. Men play their first match at age 17.5 and women at age 15.9. Women reach their highest level earlier than men, and for both men and women the peak performance tends to occur at a younger age now than in the past.

Dudink (1994) studies the birthdate distribution of successful junior tennis players in The Netherlands. He finds that the distribution is not symmetric; it is skewed, with half of the twelve- to sixteen-year-old top Dutch tennis players being born in the first three months of the year. This is due to the cut-off date for junior competition age groups, namely 1 January. Edgar and O'Donoghue (2005) confirm this season-of-birth bias, and they find a similar pattern for senior players, using grand slam data for both men and women. Notable exceptions are Aranxa Sanchez-Vicario and Richard Krajicek, both born in December and nevertheless grand slam champions.

Finally, Radicchi (2011) estimates who is the best male player ever. He considers more than 130,000 matches between 1968 and 2010 and forms a network where players are linked through matches. Jimmy Connors turns out to be the best player, justified by his extremely long, consistent, and successful career, holding a top-ten position for sixteen consecutive years (1973–1998).

6 The method of moments

To make further progress we need to know more about estimation theory. For example, we want to account for the clustering of points in matches and players, and this can only be achieved by introducing more theory. In this chapter we introduce a much-applied statistical theory: the method of moments, more precisely the generalized method of moments (GMM). The word 'moments' refers to statistical operators, such as the expectation and the variance (first encountered and defined on page 71). The method will handle the problems encountered so far, it can be extended later on, and will be used extensively in the following chapters. We begin with the simplest setup so that we can link the new results directly to what has been learned in the previous chapter. Then we gradually add new elements.

Our summary statistics are too simple

In the previous chapter all points were pooled. We distinguished between men and women, but otherwise we did not distinguish between matches or players. For example, the probability of winning a point on service was estimated by $\hat{p}$, the total number of points in the data set won on service divided by the total number of points served. This simplicity may be sufficient for some purposes, but not if we wish to dig deeper.

One problem of pooling is that all points are treated as independent and identically distributed (iid). This may be a reasonable assumption within one player within one match, and this is the content of hypothesis 1. But the hypothesis does not suggest that *all* points in the data set are iid, and of course they aren't. Points are clustered in matches and, within a match, in players. Each player has his or her own underlying probabilities and the performance of one player depends on the opponent in the match.

Thus motivated we now think of a *match* as the unit of analysis, and this necessitates a refinement of our notation. Suppose there are N matches in our data set (N is 258 for the men and 223 for the women), and consider a match between two players $\mathcal{I}$ and $\mathcal{J}$. Player $\mathcal{I}$ serves T_i points and wins a fraction f_i of them. This clusters the points served by $\mathcal{I}$. The points served by $\mathcal{J}$ are similarly clustered. Per match we observe two relative frequencies, f_i of player $\mathcal{I}$ and f_j of player $\mathcal{J}$.

As in the previous chapter, we first focus on the probability p to win a point on service. A natural alternative to the previously defined estimator $\hat{p}$, which, in contrast to $\hat{p}$, accounts for the clustering, is obtained by taking the average of f_i and f_j in a match and then average over all N matches in the sample. In formula,

$$\hat{p}_m = \frac{1}{2N} \sum (f_i + f_j),$$

where the Greek capital sigma is the mathematical symbol for the summation operator, here over all matches.

Under appropriate assumptions, the standard error accompanying $\hat{p}_m$ is

$$\text{se}_m = \frac{1}{2N} \sqrt{\sum \left(\frac{f_i(1-f_i)}{T_i} + \frac{f_j(1-f_j)}{T_j} \right)}.$$

What are these appropriate assumptions? The two expressions $f_i(1-f_i)/T_i$ and $f_j(1-f_j)/T_j$ in the above formula are the estimated variances of f_i and f_j, respectively. They resemble the expression in the old variance formula $\hat{p}(1-\hat{p})/T$ on page 71, but now they are formulated for each player separately because we only assume iid for points within one player. If we ignore (for now) the correlation between the two opponents $\mathcal{I}$ and $\mathcal{J}$, and if matches are uncorrelated, then the formula for the standard error is correct.

Table 6.1 presents three different estimates for the same overall probability p (points won on service) and also for the underlying and associated probabilities. The estimate in the column 'point' is obtained by taking the mean across all points. Regarding p, this is the estimate $\hat{p}$ of Chapter 5. The second estimate is $\hat{p}_m$, which averages matches and players. The estimate in the GMM column is also a match-based estimate and will be discussed later in this chapter (page 91).

	Men			Women		
	Point	Match		Point	Match	
	mean	mean	GMM	mean	mean	GMM
1st service (s.) in	59.4	59.4	59.4	60.8	60.5	60.5
	(0.2)	(0.3)	(0.4)	(0.3)	(0.5)	(0.7)
2nd service in	86.4	86.6	86.4	86.0	85.9	85.7
	(0.2)	(0.3)	(0.4)	(0.3)	(0.5)	(0.7)
Points won if 1st s. in	73.3	73.5	73.3	62.2	63.0	62.5
	(0.2)	(0.3)	(0.5)	(0.4)	(0.6)	(0.6)
Points won if 2nd s. in	59.4	59.6	59.4	54.1	54.2	53.5
	(0.3)	(0.5)	(0.5)	(0.5)	(0.8)	(0.9)
Points won on 1st s.	43.6	43.7	43.6	37.8	38.0	37.8
	(0.2)	(0.3)	(0.4)	(0.3)	(0.5)	(0.4)
Points won on 2nd s.	51.4	51.7	51.4	46.6	46.7	45.8
	(0.3)	(0.4)	(0.5)	(0.5)	(0.8)	(0.8)
Points won on service	64.4	64.6	64.4	56.1	56.3	55.9
	(0.2)	(0.3)	(0.3)	(0.3)	(0.5)	(0.4)
Ace	8.2	8.3	8.2	3.1	3.3	3.3
	(0.1)	(0.1)	(0.3)	(0.1)	(0.2)	(0.2)
Double fault	5.5	5.5	5.5	5.5	5.7	5.7
	(0.1)	(0.1)	(0.2)	(0.1)	(0.2)	(0.3)

Table 6.1: *Estimates of the mean of service probabilities: sample means versus generalized method of moments (GMM)*

The point-based and match-based means show that the estimates do not differ much. Both estimates thus make sense. But when we account for clustering using the match-based approach, the standard errors are always higher (for the women, much higher) than when we don't. The simple analysis of pooling all points therefore suggests a higher precision than is justified by the data, thus invalidating inference.

So far, the match analysis has accounted for one aspect of clustering, namely that each cluster of points served by one player in one match has its own winning probability. This has improved the estimation of the standard error. On the other hand, the estimator $\hat{p}_m$ no longer uses information on the number of service points T_i, because only the fractions f_i and f_j are used in the formula on page 86. This information is relevant, because a player who serves in a long match generates a more precise relative frequency f_i: more points, more information, more accurate estimates. Moreover, the estimators so far are based on the assumption that the performance of one player is unrelated to the performance of the opponent. This can't be true. A high value of f_i means not only that $\mathcal{I}$ scores many points on service against $\mathcal{J}$, but very likely that he or she is a better overall player (better returner, better in the rally), and therefore that f_j will be lower. In other words, f_i and f_j are (negatively) correlated, and so far we have not taken this into account.

One could try and derive improved versions of $\hat{p}_m$ and se_m to account for these issues. The formulas would then become more complicated, but it would be possible. Still, this would not resolve all our problems, because other issues will appear that cannot be tackled in this manner. For example, at some point we wish to incorporate observable differences between players. Some have a higher world ranking than others and we want to exploit such information. A more general approach is required. We need to know more about the theory of estimation, and the method of moments provides the appropriate framework.

The method of moments

In statistics we typically study a situation where we want to say something about a characteristic, say θ, of the population, but the only thing we have is a sample. If this characteristic (parameter) is linked to a moment, such as the mean, then we could calculate the corresponding moment in the sample, equate this to the population moment, and solve for the parameter, thus obtaining an estimate of that parameter.

For example, if we are interested in the mean income θ of all employees in The Netherlands (the population) and we have a random sample of one thousand Dutch employees, then we estimate θ by the mean income in the sample. This is the simplest application of the method of moments, estimating one parameter using one moment equation.

The method can also be used with two or more moments. If there are M population moments containing M parameters of interest, then one can equate the M corresponding sample moments to these population moments. Solving the M equations in M unknowns provides estimates of all the parameters. This is the method of moments.

If we have fewer moment restrictions than parameters, then we cannot identify the parameters. Estimation is not possible. But if we have more moment restrictions than parameters, then we have 'too much' information. This is a good thing, but in general no solution then exists that satisfies all restrictions. We have to develop a method which selects the estimate that brings us 'as close as possible' to satisfying all restrictions. This method is the *generalized* method of moments (GMM), and it contains the (ordinary) method of moments as a special case.

Thinking again of the income of Dutch employees, suppose we have two sources of income data for each person: one from the tax authority and one from the employee's bank where the monthly paycheck is received. The two averages of the two samples will not be the same, so we have two conflicting pieces of information on just one common parameter θ. GMM will then provide one overall estimate which finds the right balance between the two averages.

The (generalized) method of moments has considerable intuitive appeal and one can prove beautiful theorems concerning the asymptotic behavior of the estimators. (In the tennis setting, this concerns the hypothetical behavior of the estimators if we would have an infinite number of matches in the data set.) Of particular interest is the variance in the asymptotic distribution. It can be computed in such a way that it controls for many types of dependencies and irregularities in the data. The GMM asymptotic variance serves as an approximation to the variance in finite samples, such as our tennis data set. The standard errors that we will use are based on this robust approach and are thus reliable.

Enter Miss Marple

The method of moments is used extensively in statistics, but not only there. An example from the crime literature is Miss Jane Marple, heroine sleuth in Agatha Christie's novels. Miss Marple lived in the pretty village of St Mary Mead, a hotbed of crime, where she was involved in no fewer than sixteen murders (only counting murders in the village) over a period of some forty years. 'Dear, dear', she would say, 'I have dropped another stitch. I have been so interested in the story. A sad case, a very sad case. It reminds me of old Mr Hargraves who lived up at the Mount'. After this statement, Miss Marple explains, whilst rambling about maids, desserts and dead-and-gone Hargraves, until there, laid before them all, is the solution.

Before entering on the big stage, Miss Marple learned her trade by solving 'trivial' problems in St Mary Mead. Why a gill of picked shrimps was found where it was. What happened to the vicar's surplice. In her learning period, all 'crimes' were small and of no interest to the police, but they provided Miss Marple with a seemingly infinite number of examples of the negative side of human nature. Then, in her life as a sleuth, no crime could arise without reminding Miss Marple of some parallel incident in the history of her time. Miss Marple's acquaintances are sometimes bored by her frequent analogies to people and events from St Mary Mead, but these analogies always lead Miss Marple to a deeper realization about the true nature of a crime, and ultimately to the solution. This is Miss Marple's method and, in its essence, it is the method of moments.

How well the method works in practice is not easy to answer. It worked for Miss Marple, not only because of the method, but also because of the clever way she applied it.

Re-estimating p by the method of moments

We shall introduce the application of the method of moments to our tennis problem in three stages, corresponding to one, two, or three moment conditions. The current section uses just one moment, the mean.

In a match between players $\mathcal{I}$ and $\mathcal{J}$, the probability p_i that $\mathcal{I}$

wins a point on service against $\mathcal{J}$ is not known. We observe only the relative frequency f_i. (The same holds of course for p_j, the probability that $\mathcal{J}$ wins a service point against $\mathcal{I}$, so we do not discuss it separately.) We denote the expected value $\mathrm{E}(f_i)$ of f_i by β_i. In the current chapter we assume that β_i is constant across players and we write β_0 instead of β_i to emphasize this fact. In Chapter 7 we shall relax this assumption.

The frequency differs from the probability by random noise. Because this noise vanishes on average, the expected value of f_i equals the overall average probability p that a player wins a point on service. Hence, β_0 is simply p. The parameter β_0 is therefore an interesting parameter to estimate. Our simplest version of the GMM procedure is based on one moment condition,

$$\mathrm{E}(f_i) = \beta_0.$$

We thus have one parameter β_0 and also one moment, so that GMM boils down to the ordinary method of moments.

Employing the moment condition we obtain the results in the column headed 'GMM' in Table 6.1. The estimates are about the same as in the previous two columns, but the standard errors are again higher. In fact, they are about twice the naive standard errors of Chapter 5 using point data, as reported in the 'point' column.

The reason for the increase in standard errors when we go from left to right in the table is that we make fewer assumptions on the independence and identical distribution of the points. The point-based analysis assumes that all points are iid. The simple match-based method assumes that all points served by the same player in one match are iid and that the service points of the two players in one match are not correlated. GMM, on the other hand, does not impose these restrictions. In our case, avoiding potentially problematic assumptions has little effect on the estimates, but the precision of the estimates is more accurately estimated than before.

Men versus women revisited

Having accurate precision estimates is important, as we now demonstrate. In the previous chapter (pages 76–78), we considered hypothesis 6: the probability that the service is in is the same in the men's singles as in the women's singles. Based on the confidence intervals

$$(-2.08, -0.72) \text{ (1st service)}, \qquad (-0.42, 1.18) \text{ (2nd service)},$$

we concluded that the probability that the first service is in is not the same in the men's singles as in the women's singles, but that the probability that the second service is in may be the same.

These intervals correspond to the numbers in the column 'point' of Table 6.1. We can now repeat the analysis based on the estimates in the GMM column. Then we find

$$(-2.69, 0.53) \text{ (1st service)}, \qquad (-0.81, 2.21) \text{ (2nd service)},$$

and we see that — contrary to the results in Chapter 5 — both confidence intervals now cover zero, so that neither hypothesis can be rejected. The reason why the conclusion has changed is that the earlier standard errors were smaller than the correct (GMM) ones, making false rejections of the hypothesis more likely. Having accurate estimates of the standard errors is thus of great importance in reaching statistically sound conclusions.

This conclusion (that the probability that the service is in may be the same in the men's singles as in the women's singles) is confirmed by considering double faults. On page 80 we found a confidence interval of $(-0.29, 0.35)$. Using GMM we find the interval $(-0.84, 0.58)$. This shows again that hypotheses 6 and 7 cannot be rejected.

Beyond the mean: variation over players

So far the analysis has focussed on the mean of the service probabilities. But the probabilities p_i differ across players, and we wish to know the magnitude of these differences. Each player $\mathcal{I}$ has his or her own relative frequency f_i in a match. Because f_i is a proxy of the true probability p_i, the spread of the f_i reveals how the p_i differ across players. Or does it?

Figure 6.1 visualizes the relative frequencies. The histogram shows the number of servers that have a frequency f_i inside a given frequency interval. We already know that the average of the frequencies f_i over all players is 64.6% in the men's singles and 56.3%

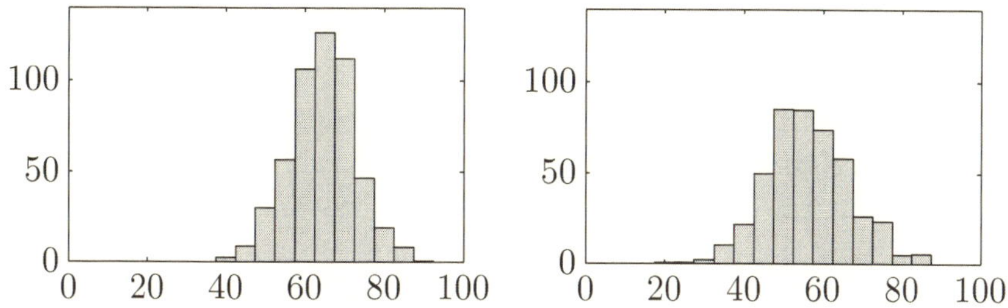

Figure 6.1: *Histogram of relative frequency of winning a point on service (men left, women right)*

in the women's singles. This is about the middle of the histograms. The histograms show that there is quite a bit of variation: there are matches where one of the players realizes a relative frequency of 80% and matches with only 40%.

Since there are more matches for the men than for the women in our data set, the area under the histogram in Figure 6.1 is larger for the men, which complicates comparisons. To transform the histograms so that they become independent of the number of matches, we plot so-called densities. A density is just the same as a histogram, except that the area under the curve is now the same for both men and women, and also that the curve is 'smoothed'.

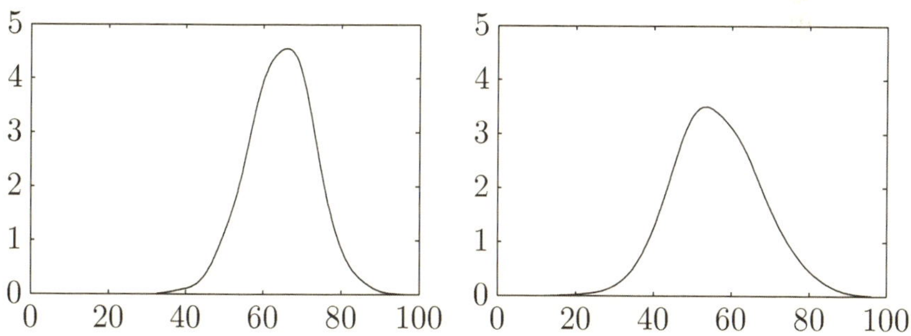

Figure 6.2: *Density of relative frequency of winning a point on service (men left, women right)*

Figure 6.2 demonstrates that the distribution of the relative frequencies is fairly symmetric around the mean, and that the spread is smaller for the men than for the women. The spread can be quantified by the standard deviation $\mathrm{sd}(f_i)$ of f_i, and in this case sd = 8.1% for the men and 10.2% for the women. The probability that we will encounter a relative frequency as high as 80% (or higher) is 3.7% for the men and 2.5% for the women. Apparently, such high relative frequencies of winning a point on service are rare, but not that rare. This is true for the men, but also for the women. Even though the mean is much lower for the women (56.3%) than for the men (64.6%), the standard deviation sd is larger for the women, so that relative frequencies far away from the mean can occur in practice.

Reliability of summary statistics: a rule of thumb

The previous section was concerned with the spread in the relative frequencies f_i across players. This is not the same as the spread in the true probabilities p_i across players, because frequencies are not the same as probabilities — they differ due to random noise. If we can use the observed spread in the frequencies to derive an indication of the spread in the true probabilities, then we obtain information about the relevance of the random noise, and thus also about how much frequencies tell us about probabilities. In tennis terminology, how informative are match (and set) summary statistics for the true performance of the player?

Hypothesis 9: *Summary statistics give a precise impression of a player's performance.*

Noise averages out in the mean, so it cannot be estimated from averages. Noise does not, however, drop out of the variance: more noise, higher variance. According to Table 5.2, each player in an average match serves 115 service points in the men's singles, and 66 points in the women's singles. Are these numbers large enough to ignore the noise and treat f_i as the true probability p_i? They are not.

To see why, consider the standard deviation 8.1% in Figure 6.2 (left panel). This standard deviation implies a variance $\text{var}(f_i)$ of $0.081^2 = 0.0066$. The variance consists of a structural part, that is, the variance of the true probabilities across players, and a remainder due to noise. For a given probability p_i and a given number of points served T_i, we know (page 86) that the variance caused by noise can be estimated by $f_i(1 - f_i)/T_i$. For the average male player we also know that $f_i = 64.6\%$ and $T_i = 115$. So the noise part of the variance is $0.646 \times (1 - 0.646)/115 = 0.0020$, which is about one-third of the total variance, and hence definitely not negligible. For the women the total variance is 0.0104 and the noise part is 0.0037, also about one-third.

The relevance of the noise becomes even stronger when we consider other probabilities, such as the probability of winning a point on 2nd service, because these rely on fewer points. We thus reject hypothesis 9. Match and particulary set summary statistics exhibit a substantial amount of noise. The fewer points are involved in computing the statistic, the larger is the noise.

We next derive a rule of thumb for interpreting summary statistics. Let f_i denote a summary statistic based on T_i points. The accuracy of this relative frequency is measured by the standard error

$$\mathrm{se} = \sqrt{\frac{f_i(1 - f_i)}{T_i}}.$$

As argued on page 71, the frequency $\pm$ twice the standard error (the confidence interval) covers the true probability with 95% probability. For example, suppose the summary statistic 'points won on service' is $f_i = 64.6\%$ when a player has served $T_i = 115$ points. The implied standard error is se = 4.5%, giving a confidence interval of 64.6%±9.0%. So, instead of saying that the probability of the player in this match was 64.6%, we should say that it lies between 55.6% and 73.6%. The player has in fact won 64.6%-points on service, but because of the presence of noise, this does not necessarily reflect his true strength in the match.

Figure 6.3 quantifies the reliability of summary statistics for various numbers of points used to compute them. The top curve applies when a summary statistic is 50%. If this is based on twenty points, then the standard error se is 11%, so that twice the standard error is 22%, as shown in the figure. The summary statistic then tells us that the player's true probability lies between 28% and 72%, not very informative. If the same summary statistic is based on a hundred points, then the curve shows that se = 5%, so that the player's true probability lies between 40% and 60%, a little more informative.

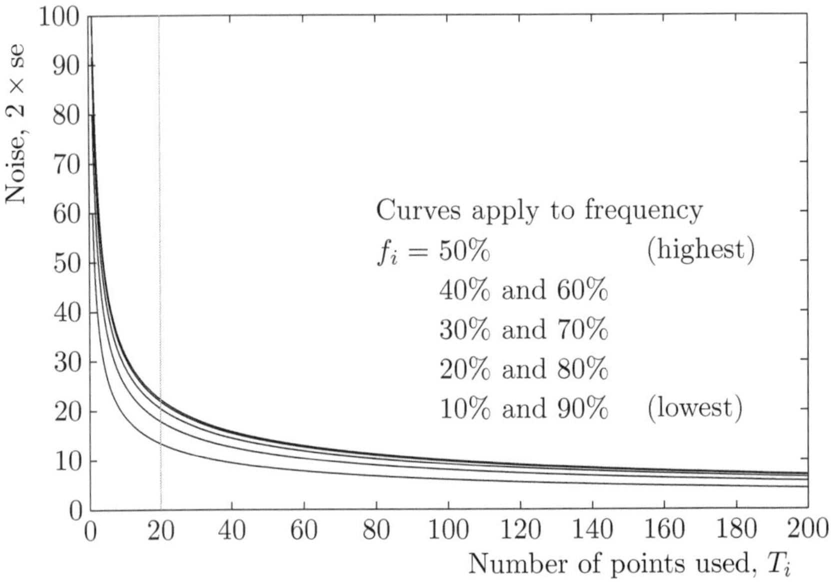

Figure 6.3: *Noise in summary statistics (measured as twice the standard error,* se*) depending on the number of points used*

Based on this figure, we derive the following rule of thumb for the reliability of summary statistics. If the number of points used to compute a summary statistic is smaller than twenty, then there is too much noise and the statistic is not informative. But if there are twenty or more points, then the true probability lies in a band of about ±10%-points around the frequency, and it may or may not be informative.

Let us reconsider the Djokovic-Nadal Australian Open 2012 final as an example. The first two sets were long and close. Djokovic lost the first set 5-7 and won the second 6-4. The third set was short and Djokovic won it 6-2, suggesting a dip in Nadal's performance. But was this dip structural or was it just random noise? Nadal's first-service percentage dropped from 71% in the first two sets to 53% in the third set, a difference of 18%-points. It would be tempting to conclude that Nadal took more risks on his first service in the third set, and that it would be better to take less risks to get the percentage up again in the fourth set. The percentages are based on ninety-two (first two sets) and thirty (third set) points, respectively.

In both cases, the number of points is larger than twenty, so that the rule of thumb applies. This means that the first-service probability in the first two sets lies between 61% and 81%, and in the third set between 43% and 63%. The statistically correct conclusion is therefore that there is no evidence that Nadal changed anything. The difference in percentages can easily be explained by chance. Hence, to conclude from these percentages that Nadal changed his service strategy in the third set is not justified.

Filtering out the noise

The magnitude of noise in observed frequencies implies that we need to filter out this noise if we wish to say something meaningful about the true probabilities. One moment condition does then not suffice. The first moment condition, derived on page 91, is still valid, but we need more.

The first moment condition reflects the fact that the frequency f_i differs from the overall average probability β_0 by an unexplained part with zero expectation. We now split this unexplained part into two elements: one related to the true probability p_i of player $\mathcal{I}$ and one reflecting noise.

In a match of $\mathcal{I}$ against $\mathcal{J}$, the noise (denoted by ϕ_i, the Greek letter phi) is the difference between the relative frequency f_i and the probability p_i, formally

$$f_i = p_i + \phi_i.$$

This is one unexplained element. The other unexplained element (denoted by π_i) captures the fact that the probability p_i will vary over players (heterogeneity).

Because we still focus on the overall average β_0, the sources of variation in p_i are all contained in π_i. Thus, π_i contains everything that makes player $\mathcal{I}$ different from the average player: his or her 'quality', form of the day, small injuries, or fear of the current opponent. This gives

$$p_i = \beta_0 + \pi_i.$$

Putting the two unexplained elements together we obtain

$$f_i = \beta_0 + \pi_i + \phi_i.$$

Hence, the unexplained part of f_i is indeed the sum of something that captures the variation in p_i and pure noise.

Because the unexplained part has zero expectation, the same holds for $\pi_i + \phi_i$. Without loss of generality we may normalize the individual expectations,

$$\mathrm{E}(\pi_i) = 0, \qquad \mathrm{E}(\phi_i) = 0,$$

so that both vanish 'on average'.

We have introduced two new variables, π_i and ϕ_i, but without additional information there is no way to disentangle them. To resolve this problem, we employ the lesson from the previous section that the spread of the frequencies contains information about the spread of the probabilities (that is, the variance of π_i), once we correct for the spread originating from the noise ϕ_i.

The random noise ϕ_i will be uncorrelated with π_i, so that

$$\mathrm{var}(f_i) = \mathrm{var}(\pi_i + \phi_i) = \mathrm{var}(\pi_i) + \mathrm{var}(\phi_i).$$

The variance of π_i is of particular interest, as it captures structural variation across players. It is unknown and we denote this unknown quantity by σ^2:

$$\mathrm{var}(\pi_i) = \sigma^2.$$

Regarding the variance of ϕ_i we again exploit the fact that, for given p_i, the variance is given by $p_i(1-p_i)/T_i$ if we assume that points within a player are iid. We do not know p_i, but we can 'average out' over all players. This gives

$$\mathrm{var}(\phi_i) = \mathrm{E}\left(\frac{p_i(1-p_i)}{T_i}\right) = \mathrm{E}\left(\frac{f_i(1-f_i)}{T_i - 1}\right),$$

where the second equality is derived from the two expressions

$$\mathrm{E}(f_i) = \mathrm{E}(p_i)$$

and

$$\mathrm{E}(f_i^2) = \mathrm{E}\left(\frac{p_i(1-p_i)}{T_i}\right) + \mathrm{E}(p_i^2).$$

These results allow us to formulate the next stage, based on two moment conditions.

Noise-free variation over players

The two moment conditions are

$$\mathrm{E}(f_i) = \beta_0$$

and

$$\mathrm{var}(f_i) = \sigma^2 + \mathrm{E}\left(\frac{f_i(1 - f_i)}{T_i - 1}\right).$$

Both moments link features of what we can proxy from the data (the two expectations and the variance) to parameters (β_0 and σ), which is the essence of GMM estimation. The two moment conditions give us two equations in two unknown parameters. Solving the equations produces the estimates $\hat{\beta}_0$ and $\hat{\sigma}$.

The estimates for β_0 (the location of the probabilities) are the same as before. The gain is that we now have estimates for σ (the spread of the probabilities, free of noise) as well. Table 6.2 presents these estimates (GMM) and, for comparison, also the standard deviations (sd, not free of noise) of the frequencies.

	Men		Women	
	sd	GMM	sd	GMM
1st service	6.9*	5.0*	9.4*	7.0*
2nd service in	6.2*	3.3*	8.8*	5.3*
Points won if 1st service in	9.1*	6.7*	11.8*	7.6*
Points won if 2nd service in	10.9*	6.1*	14.8*	8.9*
Points won on 1st service	7.5*	5.3*	8.8*	5.2*
Points won on 2nd service	10.5*	6.0*	13.9*	8.6*
Points won on service	8.1*	5.9*	10.2*	6.7*
Ace	6.2*	5.3*	3.2*	2.1*
Double fault	2.9*	1.8*	3.9*	2.5*

Table 6.2: *Estimates of the variation in service probabilities across players (σ): standard deviation (sd) versus generalized method of moments (GMM)*

In this and most of the following tables we do not present standard errors. Instead we indicate with * that an estimate is significant and with ° that it is not significant. This provides an easy way of testing hypotheses, as explained on pages 77 and 78. For example, Table 6.2 shows that all spread parameter estimates are significantly greater than zero, so players are heterogeneous — not surprisingly. If an estimate lacks either superscript, then the question of significance is not meaningful.

The estimated value of σ is substantially lower than the standard deviation (sd) of the relative frequencies. This confirms that the noise constitutes a large part of these relative frequencies, and that it has therefore been useful to filter it out. It also confirms our previous conclusion regarding hypothesis 9 that one has to be careful when interpreting summary statistics.

When we compare men to women, we see that the standard deviation is higher for the women. This can be due to more spread in the probabilities, but it can also be due to more noise. The latter explanation makes sense, because women serve fewer points than men, leading to more noisy frequencies. The new insight from the GMM estimates is that the first explanation also appears to be true: σ tends to be higher for the women. One possible explanation is that the quality differences for the women are larger than for the men, so that there are more matches where player $\mathcal{I}$ wins many points (high p_i) and player $\mathcal{J}$, her opponent, wins few points (low p_j), thus causing a higher spread.

Correlation between opponents

Two probabilities, p_i and p_j, govern one match. We expect these two probabilities to be correlated because of two facts. First, how well one player serves is correlated with how well he or she returns (and performs in the rally). A good player will, not always but on average, serve well *and* return well *and* play well in rallies. Second, a player who returns well (and plays well in rallies) makes it more difficult for the opponent to perform well on service. Both facts together imply that we expect the correlation to be negative: a high p_i will be associated with a low p_j.

The magnitude of this correlation is of interest and we want to estimate it. Thus we introduce a parameter ρ for the correlation between p_i and p_j. This is also the correlation between f_i and f_j, because the noise ϕ_i is uncorrelated with the opponent's f_j. If $\rho = 0$ then there is no correlation. We expect however that $\rho < 0$.

These considerations lead to a third moment condition, to be added to the existing two. We present this new moment not in terms of the correlation itself, but in terms of the 'covariance', defined by

$$\operatorname{cov}(f_i, f_j) = \operatorname{E}\left(f_i - \operatorname{E}(f_i)\right)\left(f_j - \operatorname{E}(f_j)\right).$$

The resulting moments are now:

$$\begin{aligned}
\operatorname{E}(f_i) &= \beta_0, \\
\operatorname{var}(f_i) &= \sigma^2 + \operatorname{E}\left(\frac{f_i(1-f_i)}{T_i - 1}\right), \\
\operatorname{cov}(f_i, f_j) &= \rho\sigma^2.
\end{aligned}$$

Solving these three equations in three unknowns (β_0, σ, and ρ) gives us the estimates $\hat{\beta}_0$, $\hat{\sigma}$, and $\hat{\rho}$.

	Men GMM	Women GMM
1st service in	-0.03°	0.08°
2nd service in	0.09°	-0.10°
Points won if 1st service in	-0.19°	-0.78^*
Points won if 2nd service in	-0.55^*	-0.59^*
Points won on 1st service	-0.36^*	-0.83^*
Points won on 2nd service	-0.63^*	-0.63^*
Points won on service	-0.51^*	-0.92^*
Ace	-0.11°	-0.76^*
Double fault	-0.08°	0.04°

Table 6.3: *GMM estimates of the correlation (ρ) in service probabilities between opponents*

The estimates of β_0 and σ are the same as in the case with two moments, but the estimates of the correlation ρ in Table 6.3 are new. The probability of winning a point on service is clearly negatively correlated between two opponents, as the estimated correlations are -0.51 (men) and -0.92 (women). Both are significantly negative, as indicated by the *. For the four most closely related probabilities, points won if 1st (2nd) service in and points won on 1st (2nd) service, the estimates are also negative, as expected.

For the remaining probabilities, 1st (2nd) service in, ace, and double fault, some correlations are positive, while others are negative, and all estimates but one are insignificant. This insignificance also corresponds to intuition: it is difficult to argue why these remaining probabilities should be correlated at all.

The most intriguing aspect in Table 6.3 is that the negative correlations for the women are stronger than for the men. For example, for the probability of points won on service — arguably the most important probability to estimate — we find $\hat{\rho} = -0.92$ for the women and $\hat{\rho} = -0.51$ for the men, and the difference is significantly different from zero. Why? If $\mathcal{I}$ is a good rally player, then he or she will perform well in his or her own service games (high p_i), but also when he or she receives, that is when $\mathcal{J}$ serves (low p_j). This implies a negative correlation. This negative correlation is stronger for women than for men, because in women's tennis rallies are more important in deciding the winner of a point.

Why bother?

In this chapter we introduced a method of estimation, the (generalized) method of moments. This method has considerable intuitive appeal and the resulting estimates have beautiful statistical properties. But do we really need this machinery? We shall see, in all subsequent chapters, that we do need it. The current chapter has already shown — step by step — that there is a point where taking averages and providing sample statistics does not suffice if we value correct reasoning and credible conclusions. We need proper statistics — not statistics in the sense of tables with averages, but statistics in the sense of mathematical statistics: the theory of estimation, testing, and inference.

Further reading

A complete explanation of the method of moments can be found in Hall (2005). Miss Marple's connection with the method of moments has, as far as we know, not been studied before, but most other aspects of her life have been described in Hart's (1997) masterly and amusing biography.

Pollard *et al.* (2009) discuss how players (and perhaps coaches) could exploit the information in summary statistics during a match. They suggest using the summary statistics of set one, say, to determine the service strategy for the second set. For example, if a player was particularly successful with a high-risk service in the first set, the player could improve his or her performance in the second set by making the second service riskier. The crucial assumption here is that the first-set frequencies are good proxies for the underlying probabilities. The rule of thumb on page 96 could be used to get an idea about whether the performance improvement computed from the frequencies is sufficiently large to be used as a motivation for changing strategy. In Chapter 9 we return to the question of whether and to what extent summary statistics are useful for service strategy.

At several places in this book we have compared men to women. So far, we have seen that the service dominance is larger for the men than for the women (page 17), that there are more upsets in the men's singles in line with the fact that the quality differences for the men are smaller (pages 24 and 100), that the most important point in a game in an average men's match is 30-40 for both men and women, but the second most important point is 15-40 for the men and 30-30 for the women (page 51), that the probabilities of first- and second-service in and double fault are remarkably similar (page 92), and that correlation is stronger for women than for men (page 102). Later in the book we will provide further comparisons, such as whether men's tennis is more competitive than women's tennis (page 117), how tennis professionals win points on their first services (page 121), how efficiently they serve (page 149), the quality differences between top and weaker players (page 165), the performance at important points (Chapter 11), and the possible influence of momentum (Chapter 12).

Some authors use tennis data to address other gender differences. Coate and Robbins (2001) ask whether top-250 male tennis professionals are more dedicated to their careers than the females, but they find no evidence of this. Wozniak (2012) studies whether participating in a tournament depends on performance at recent tournaments, and reports that women compete more frequently than men after a successful recent performance.

7 Quality

The success of a player depends on his or her ability, on the opponent, and on unobserved factors such as form of the day and luck. How can the quality of a player be best measured? By his or her position on the world rankings? Or should a transformation be applied to the rankings? If so, how? And is the ranking of top players a good indicator of their quality in the first few rounds, or must they grow into the tournament? These are the questions addressed in this chapter.

Observable variation over players

In the previous chapter we modeled the unobserved probability p_i — the probability that player $\mathcal{I}$ will win a point on service in a match against player $\mathcal{J}$. We concentrated on the mean value of this probability across all players, which we called β_0. Players are not all alike, so p_i deviates from this overall value, and we accounted for these deviations via the heterogeneity variable π_i.

Some of the heterogeneity is observable, some is not. Instead of writing $p_i = \beta_0 + \pi_i$, as in the previous chapter, we now write

$$p_i = \beta_i + \pi_i,$$

where β_i and π_i represent observable and unobservable heterogeneity, respectively. Observable heterogeneity is caused, for example, by the fact that players have a different position on the world rankings, while unobservable heterogeneity contains form of the day, fear of a specific opponent, and other things on which data are unavailable. If we can account for observable information, then we reduce the unexplained player heterogeneity π_i, most likely obtain more precise estimates of the parameters, and also estimate new features of tennis. We now explain how to account for observables in β_i and estimate their impact.

	Sd-Sd	Sd-NSd	NSd-Sd	NSd-NSd	Total
Men	67.0	69.3	61.1	63.7	64.4
	(0.6)	(0.4)	(0.4)	(0.4)	(0.2)
Women	56.9	62.9	50.0	55.8	56.1
	(0.8)	(0.5)	(0.5)	(0.7)	(0.3)

Table 7.1: *Percentage of points won on service, seeded (Sd) and non-seeded (NSd) players*

Table 7.1 gives a first indication of quality differences by separating the matches in four categories: seeded against seeded (Sd-Sd), seeded against non-seeded (Sd-NSd or NSd-Sd), and non-seeded against non-seeded (NSd-NSd). The difference between Sd-NSd and NSd-Sd is that in Sd-NSd we consider points where the seeded player serves, while in NSd-Sd the non-seeded player serves.

A seeded player in the men's singles achieves 67.0% service point success against another seeded player. If he plays against a non-seeded player then he has (of course) a higher service success: 69.3%. A non-seeded player in the men's singles achieves 63.7% service success against another non-seeded player, but if he plays against a seeded player then his service success is reduced to 61.1%. If two seeds meet, then, for both players, their service success (67.0%) is higher than when two non-seeds meet (63.7%), which suggests that not only the quality difference is important but that the quality sum may also matter.

The conclusions are similar in the women's singles. All service success percentages are lower and the discrepancy between Sd-Sd and NSd-NSd is smaller than in the men's singles, suggesting that quality sum may be relevant, but less so than in the men's singles.

This first indication suggests that the deterministic component β_i depends on the quality q_i of player $\mathcal{I}$ and the quality q_j of player $\mathcal{J}$, more conveniently on the quality difference and the quality sum, as follows:

$$\beta_i = \beta_0 + \beta_-(q_i - q_j) + \beta_+(q_i + q_j).$$

In the unlikely event where both new parameters β_- and β_+ are zero, our model specializes to the model of Chapter 6. Because all unobservable determinants are contained in π_i, the variables relevant to q_i are all observable, so q_i actually represents *observed* quality. We don't know yet what these variables are, so our next question is how to specify q_i.

Ranking

Every Monday, with few exceptions, the Association of Tennis Professionals (ATP, for the men) and the Women's Tennis Association (WTA, for the women) publish an updated list of the players' ranking points and rankings based on their performance over the past fifty-two weeks. The method of calculating the ranking points is somewhat complex, but the rankings $(1, 2, \dots)$ are simply the order of the players according to the ranking points. The two lists are also published just before Wimbledon, and the seedings are based (almost but not entirely) on these lists. We denote the rankings as $rank_i$ for player $\mathcal{I}$ and $rank_j$ for player $\mathcal{J}$. The lowest-ranked player in the tournament may have $rank_i$ as high as 500, even though only 128 players take part.

A natural first attempt in specifying q_i is to equate quality with ranking: $q_i = rank_i$. In that case, β_i and the winning probability p_i would depend linearly on the difference between $rank_i$ and $rank_j$. This is not satisfactory, because quality in sports is a pyramid: the difference in strength between numbers 1 and 16 is greater than between 101 and 116.

Hypothesis 10: *Quality is a pyramid.*

If quality is indeed a pyramid, then going down the pyramid taking equal steps of quality reduction, the number of players involved increases at each step. To see whether this is true we first consider the ranking points.

The 2012 end-of-year lists show that Novak Djokovic (12,920 points) and Victoria Azarenka (10,595 points) hold the number-one positions at the end of 2012. The same two lists show that there are five players in the men's singles (three in the women's singles) who have between 2000 and 2500 points, seven (thirteen) between 1500 and 2000 points, seventeen (thirty-two) between 1000 and 1500 points, and seventy-three (seventy-eight) between 500 and 1000 points. So, going down from the top by steps of 500 points each, involves more and more players.

This suggests a pyramid, but there are some problems associated with using ranking points as a quality measure. Not only does the method of computing ranking points change regularly and differ between men and women, but more importantly the ranking points are artificial creations, not directly related to the players' true qualities. Equal drops in ranking points are therefore not the same as equal steps of quality reduction, and hence we cannot conclude from the numerical exercise above that quality is a pyramid. The rankings are more robust, and we therefore wish to define quality as a function of the rankings, not of the ranking points.

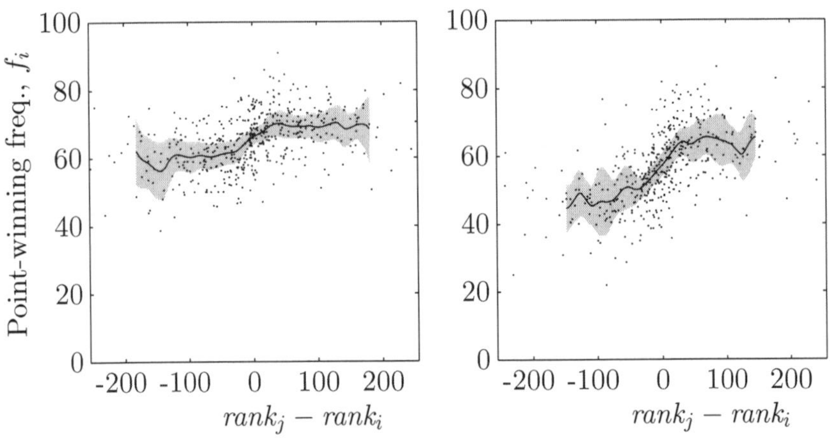

Figure 7.1: *Percentage of winning a point on service as a function of the difference in rankings (men left, women right)*

Thus motivated, let us use the rankings, rather than the ranking points, to study whether quality is really a pyramid. We have data on f_i, so we can plot f_i against $rank_j - rank_i$. (Remember that a positive value of $rank_j - rank_i$ means that $\mathcal{I}$ is a better player than $\mathcal{J}$.) The dots in Figure 7.1 represent the combinations of f_i

and $rank_j - rank_i$ for all matches (two players per match). The vertical dispersion in the scatter plot is substantial, reflecting the two unexplained elements π_i (unobserved part of p_i) and ϕ_i (pure noise).

To summarize the dots in some interpretable curve, we apply a statistical technique called 'non-parametric' mean regression. For a given ranking difference $rank_j - rank_i$, say fifty, we take the mean over the f_i for which the corresponding $rank_j - rank_i$ is around fifty. This averages out the unexplained influences on f_i to some extent, so that the impact of $rank_j - rank_i$ on β_i and thereby on p_i remains. We thus obtain an estimate of this impact, but only for $rank_j - rank_i$ around fifty. We now repeat this exercise for many values of the ranking difference, and this creates a curve of estimates. This is the essence of non-parametric mean regression. The technique does not impose linearity or any other functional form. The resulting form of the curve is therefore generated by the data and not by underlying assumptions, and this is useful if we wish to learn about the functional form of the relationship.

The solid curves in Figure 7.1, one for the men and one for the women, show that p_i depends positively on $rank_j - rank_i$, as expected. More important is the shape of the dependence. Apparently the relation is non-linear and has the form of an S-curve.

Let us now see how we can use Figure 7.1 to decide whether or not quality is a pyramid. We first keep $rank_i$ constant and vary $rank_j$. We start from $rank_j - rank_i = 0$ (the middle of the figure), so that players have equal ranking. When we increase $rank_j$ by one, then p_i increases and with each further unit increase in $rank_j$ the increase in p_i becomes smaller. At about $rank_j - rank_i = 30$ the curve can no longer be distinguished from a flat curve, so an increase from 30 to 31 hardly increases p_i any more.

Next, we keep $rank_j$ constant and vary $rank_i$. A similar flattening occurs. Starting again from $rank_j - rank_i = 0$, a unit increase in $rank_i$ will reduce p_i and with each further unit increase in $rank_i$ the reduction in p_i becomes smaller. Put differently, a given reduction in p_i involves larger and larger increments in $rank_i$ and thus more and more players. This is precisely what we mean by a pyramid. So, Figure 7.1 supports hypothesis 10, and we conclude that quality in tennis — and most likely also in other sports — is indeed a pyramid.

We would like to transform the ranking in such a way that the S-shape is removed, so that β_i depends linearly on the transformed ranking. We achieve this, based on the idea of a pyramid, by introducing the idea of 'expected round': 8 for a player with $rank_i = 1$ who is expected to reach the final (round 7) and win, 7 for a player with $rank_i = 2$ who is expected to reach the final and lose, 6 for players with $rank_i$ 3 or 4 who are expected to lose in round 6, 5 for players with $rank_i$ 5 to 8 who are expected to lose in round 5, and so on. When we walk down the pyramid, more and more players get the same quality indicator, capturing the flattening of quality differences represented by the idea of a pyramid. A problem with the expected round, however, is that it does not distinguish between, for example, players ranked 9 to 16, because all of them are expected to lose in round 4.

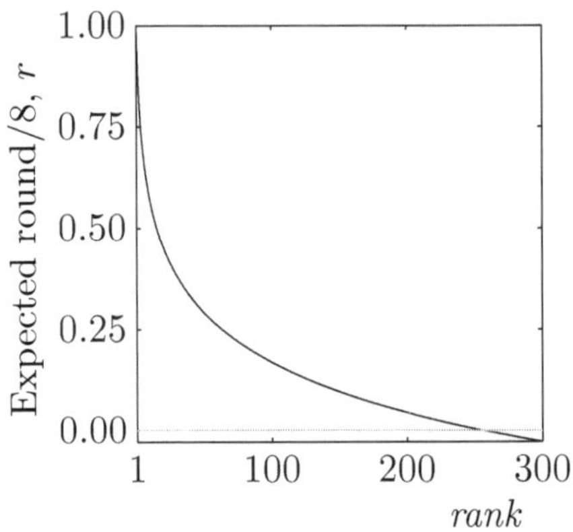

Figure 7.2: *Expected round variable r as a function of the ranking*

A smooth measure of expected round is provided by the formula

$$\textit{expected round}_i = 8 - \log_2(\textit{rank}_i),$$

where $\log_2(x)$ denotes the logarithm of x with base 2. It will be convenient to scale this formula such that the maximum is 1 rather than 8. This then leads to the transformed ranking variable

$$r_i = \frac{\textit{expected round}_i}{8} = 1 - \frac{\log_2(\textit{rank}_i)}{8},$$

which is plotted in Figure 7.2. If $rank_i = 1$ then $r_i = 1$ (the maximum) reflecting that we expect the world number one to win the tournament. If $rank_i = 9$ then $r_i = 0.60$, and if $rank_i = 16$ then $r_i = 0.50$. The value $r_i = 0.50$ can be interpreted as reaching '50% of the tournament' (round 4 of 8). For players with $rank_i > 128$, we have $r_i < 1/8$, which means that the player was not expected to participate in the tournament based on his or her ranking. Players with $rank_i > 256$ will have a negative r_i, but this causes no problems.

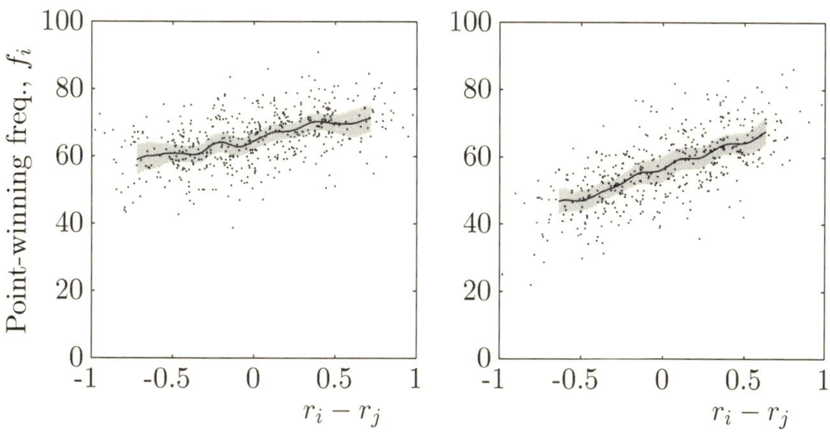

Figure 7.3: *Percentage of winning a point on service as a function of the difference in expected rounds (men left, women right)*

Can the (smoothed) expected-round idea explain the *S*-curve suggested by the data? To test this we perform another nonparametric mean regression, this time on $r_i - r_j$ instead of on $rank_j - rank_i$. Figure 7.3 shows that the dependence of β_i and thus p_i on $r_i - r_j$ is now close to linear. The expected-round transformation implies that constant step-by-step reductions in p_i correspond to constant reductions in r_i, which involve more and more players. Hence, the transformation fully captures the idea of a pyramid and converts $rank_i$ into units of quality. For example, the quality difference between the numbers 1 ($r_i = 1$) and 16 ($r_i = 0.5$) is twenty times the quality difference between the numbers 101 ($r_i = 0.168$) and 116 ($r_i = 0.143$).

With this knowledge we now specify

$$q_i = r_i,$$

in which case β_i can be written as

$$\beta_i = \beta_0 + \beta_-(r_i - r_j) + \beta_+(r_i + r_j).$$

This would suffice if rankings were the only determinant of p_i that we can observe and measure. But maybe there is more.

Round, bonus, and malus

Perhaps the performance of a player depends, in addition to the ranking of the player and his or her opponent, on the round they are playing in. A low-ranked player who has managed to progress in the tournament has apparently higher quality than his or her ranking indicates, and a top player may not play his or her best tennis in the early rounds. The latter statement in particular is often heard and therefore worth investigating.

Hypothesis 11: *Top players must grow into the tournament.*

To analyze both possibilities we distinguish between 'bonus',

$$bonus_i = \max\left(round_i/8 - r_i,\ 0\right),$$

and 'malus',

$$malus_i = \min\left(round_i/8 - r_i,\ 0\right).$$

If $round_i/8 > r_i$, then $\mathcal{I}$ has progressed further in the tournament than could have been expected on the basis of ranking. There is a positive bonus and no malus. However, if $round_i/8 < r_i$, then $\mathcal{I}$ is a top player in an early round, with a negative malus and no bonus.

We specify

$$q_i = r_i + \alpha_b bonus_i + \alpha_m malus_i.$$

For given values of the parameters α_b and α_m, the quality q_i depends only on r_i and $round_i$, and Figure 7.4 plots this relationship for $\alpha_b = 0.8$ and $\alpha_m = 0$. The figure contains seven partly overlapping kinked lines, one for each round. Consider a player $\mathcal{I}$ who

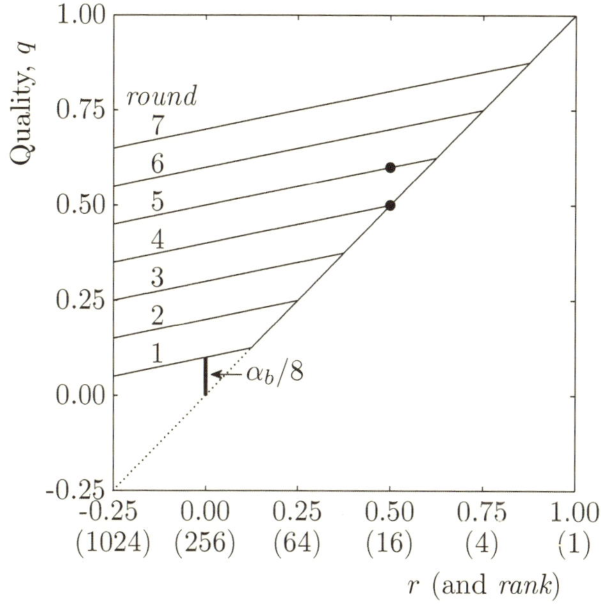

Figure 7.4: *Quality as a function of round and expected round,* $\alpha_b = 0.8$, $\alpha_m = 0$

is number 16 on the ATP world rankings, so that $r_i = 0.50$. This player is expected to reach round 4, but not round 5. In round 1, $bonus_i = 0$ but $malus_i$ is negative. However, since $\alpha_m = 0$ in the figure, his quality is unaffected so that $q_i = r_i$, as indicated by the lower of the two dots. If $\mathcal{I}$ wins and moves to the second round, this is no outstanding achievement for him, so there is still no bonus, q_i remains r_i, and the same dot applies. If he continues winning and reaches the quarter-finals (round 5), then he performs better than his ranking suggests and $bonus_i = 1/8$. His quality indicator is increased by $0.8/8 = 0.1$ and becomes $q_i = 0.6$, represented by the upper dot. If $\mathcal{I}$ reaches the semi-finals, then his q_i moves up to 0.7; if he reaches the final, to 0.8.

Let us now return to hypothesis 11: top players must grow into the tournament. If $\alpha_m = 0$ then the quality of a top player in the early rounds is as expected, but if $\alpha_m > 0$ then the quality is lower than expected. Therefore we can test hypothesis 11 by testing $\alpha_m = 0$ versus $\alpha_m > 0$. Performing this test we find that the hypothesis $\alpha_m = 0$ cannot be rejected, which means that we have no evidence supporting hypothesis 11. One explanation could be that professional tennis is so competitive that top players can not and do not relax in the early rounds.

We therefore impose $\alpha_m = 0$. But we do not (yet) impose $\alpha_b = 0$. We estimate α_b by $\hat{\alpha}_b = 0.7$ (0.2) for the men and $\hat{\alpha}_b = 0.8$ (0.2) for the women. The player ranked 16 in the men's singles achieves $q_i = 0.59$ when he reaches the quarter-finals, so that he is expected to perform as if he were ranked number 10. A woman player ranked 16 achieves $q_i = 0.60$ when she reaches the quarter-finals, and is expected to be of the same quality as the world's number 9.

Significance, relevance, and sensitivity

We set $\alpha_m = 0$, because its estimate is not significantly different from zero. We now also set $\alpha_b = 0$, in spite of the fact that its estimate is significant. With both α_m and α_b equal to zero, the quality q_i depends only on the rank: $q_i = r_i$.

Surely, this is a bit strange. The estimate of α_b is significantly different from zero, and still we set the parameter equal to zero. How can we justify such a modeling decision? For this we need to understand the difference between significance and relevance.

Suppose we consider a model where the variable to be explained, say h, depends on u and a number of other variables:

$$h = \theta u + \text{other things.}$$

We don't know the value of the parameter θ, which we assume not to be zero, but we can estimate it when appropriate data are available. The more (and better) data we have, the more accurately we can estimate θ, and the smaller will its standard error be. When the standard error is very small, then the hypothesis that $\theta = 0$ will almost certainly be rejected. Put differently, the estimate $\hat{\theta}$ will be 'significant', as explained on page 77.

It is perfectly possible, however, that the *impact* of θu on h is small compared to the impact of the other variables in the model, even though $\hat{\theta}$ is significant. The variable u is then not 'relevant'. We want to avoid variables that are not relevant, because estimating the associated parameters adds uncertainty and this makes the standard errors of the other estimates larger, which is undesirable.

What we really want to know is not so much whether the estimate of a parameter is significant or not, but whether the answer to the question under investigation changes when we set the parameter equal to zero. If the answer changes, then our research question is sensitive to this parameter; if not, it is not sensitive.

Significance and sensitivity are two different concepts, and they are essentially unrelated. Knowledge of one tells you very little about the other. A parameter estimate may be significant but the answer to our research question may not be sensitive to setting the parameter equal to zero, and vice versa.

In our case, the estimate of α_m is not significant, while the estimate of α_b is significant. Neither parameter is sufficiently relevant. Therefore we set both parameters equal to zero and include neither malus nor bonus in our model.

The complete model

This completes our model. Let us briefly summarize it. The relative frequency f_i and the probability p_i are related through noise ϕ_i:

$$f_i = p_i + \phi_i.$$

The relative frequency can be observed from the data, but the probability cannot be observed and needs to be modeled. We assume that it depends on deterministic components β_i and random components π_i:

$$p_i = \beta_i + \pi_i,$$

where we have specified β_i as

$$\beta_i = \beta_0 + \beta_-(r_i - r_j) + \beta_+(r_i + r_j).$$

Combining terms we find

$$f_i = \beta_0 + \beta_-(r_i - r_j) + \beta_+(r_i + r_j) + \pi_i + \phi_i.$$

From now on we shall usually write 'ranking' when we mean 'transformed ranking', unless there is a possibility of confusion.

For each match the average of the two ranking differences is zero, so the average $r_i - r_j$ in the sample is also zero. We subtract the average ranking sum from $r_i + r_j$ ('centering'), so that the ranking sum also has mean zero. This gives $\beta_i = \beta_0$ for the average match, so that β_0 keeps its interpretation as the overall average probability. The centering is only introduced for convenience of interpretation; it does not affect the results.

The error term is $\pi_i + \phi_i$, and we need assumptions on its behavior in order to derive moment conditions for estimation, as in the previous chapter. The error term still has zero expectation, so that the original moment condition $\mathrm{E}(f_i) = \beta_i$ still applies. In contrast to the previous chapter, however, we now have three parameters in β_i, namely β_0, β_-, and β_+, instead of just β_0, and we cannot solve all three parameters from a single moment condition. Fortunately, the ranking variables r_i and r_j are uncorrelated with the error, and this yields two additional moment conditions:

$$\mathrm{E}\left((f_i - \beta_i)(r_i - r_j)\right) = 0, \qquad \mathrm{E}\left((f_i - \beta_i)(r_i + r_j)\right) = 0.$$

Under suitable assumptions, discussed in the previous chapter, we then arrive at the five moment conditions,

$$\begin{aligned}
\mathrm{E}(f_i - \beta_i) &= 0, \\
\mathrm{E}\left((f_i - \beta_i)(r_i - r_j)\right) &= 0, \\
\mathrm{E}\left((f_i - \beta_i)(r_i + r_j)\right) &= 0, \\
\mathrm{var}(f_i) &= \sigma^2 + \mathrm{E}\left(\frac{f_i(1 - f_i)}{T_i - 1}\right), \\
\mathrm{cov}(f_i, f_j) &= \rho\sigma^2.
\end{aligned}$$

These five conditions allow us to estimate the unknown parameters β_0, β_-, β_+, σ, and ρ and compute their standard errors.

Winning a point on service

We first present the estimation results for p_i, the probability of winning a point on service. In the first and third rows of Table 7.2 we estimate the restricted model where $\beta_- = \beta_+ = 0$, so that quality $\beta_i = \beta_0$ is constant for all players. This is the model estimated in Chapter 6, and the estimates (64.4% for the men and 55.9% for the women) correspond to the estimates presented in Table 6.1, column GMM. The estimates of the heterogeneity parameter σ and the correlation parameter ρ also agree with the estimates in Chapter 6.

	β_0 mean	β_- $r_i - r_j$	β_+ $r_i + r_j$	σ heterog.	ρ correl.
Men					
Not using ranking	64.4	–	–	5.9*	−0.51*
Using ranking	64.9	8.0*	3.1*	5.0*	−0.44*
Women					
Not using ranking	55.9	–	–	6.7*	−0.92*
Using ranking	56.3	15.9*	1.8°	4.2*	−0.79*

Table 7.2: *Explaining the probability of winning a point on service*

In the second and fourth rows we estimate the complete model. Taking quality into account hardly affects the estimates of β_0, which is consistent with the fact that β_0 still reflects the overall average probability of winning a point. The effect of quality difference ($r_i - r_j$) on p_i is much larger than the effect of quality sum ($r_i + r_j$), as expected.

Hypothesis 12: *Men's tennis is more competitive than women's tennis.*

This issue was addressed in Chapter 2, where we concluded that there are more upsets (top-sixteen seeds not reaching the last sixteen) for the men than for the women, in line with the hypothesis. Table 7.2 provides new evidence based on two improvements: we work at point level rather than at match level, so that the analysis now takes account of the difference between best-of-five versus best-of-three matches; and we use rankings, which is a better indicator of quality than the earlier distinction between seeds and non-seeds.

The estimated β_- is smaller for the men (8.0) than for the women (15.9), and significantly so. Apparently, the ranking is more informative for women than for men, which means that the difference in strength between top and lower-ranked players is greater in the women's singles than in the men's singles, again supporting hypothesis 12.

We notice the possibility of confusion in the terms 'lower rank' and 'higher rank'. The best player has world ranking 1, but in common language (and also in tennis terminology) such a player has a high rather than a low ranking. We follow common language and emphasize that here and in the following a lower-ranked player is a player with a lower r_i-value than a higher-ranked player, that is, a player of lower quality.

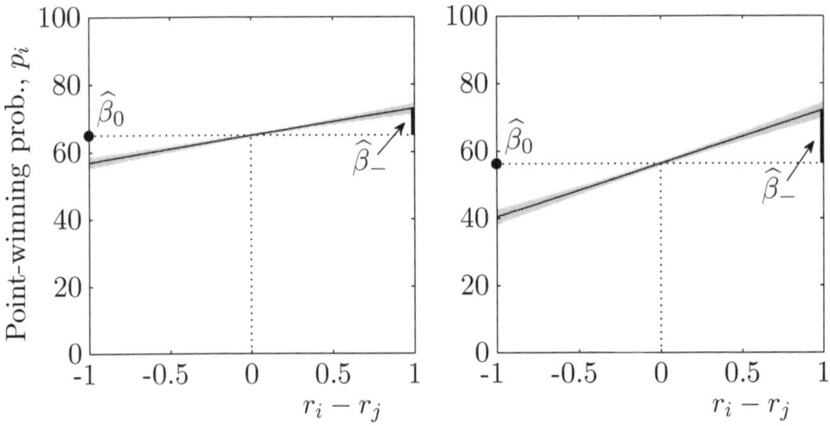

Figure 7.5: *Probability of winning a point on service as a function of expected-round difference (men left, women right)*

We plot the dependence of p_i on quality difference in Figure 7.5 for a player having zero unobserved quality π_i, keeping quality sum at its average value. At the figure's extreme we have $r_i - r_j = 1$ and hence the probability of winning a point for the men is $\hat{\beta}_0 + \hat{\beta}_- =$ 72.9% for player $\mathcal{I}$ and $\hat{\beta}_0 - \hat{\beta}_- =$ 56.9% for player $\mathcal{J}$. For the women, the probability of winning a point is $\hat{\beta}_0 + \hat{\beta}_- =$ 72.2% for player $\mathcal{I}$ and $\hat{\beta}_0 - \hat{\beta}_- =$ 40.4% for player $\mathcal{J}$. The shadow around the estimated line represents estimation uncertainty, implied by the standard errors of the estimated parameters. The estimation uncertainty turns out to be small, much smaller than the heterogeneity σ.

Table 7.2 also shows that there is more unobserved heterogeneity ($\hat{\sigma} = 5.0$) for the men than for the women ($\hat{\sigma} = 4.2$), which is counterintuitive and therefore worth noting. It is counterintuitive because the men's game is more competitive than the women's game, and one would therefore expect less rather than more heterogeneity (more rather than less homogeneity). Indeed, when we ignore the quality effect, we do find less heterogeneity ($\hat{\sigma} = 5.9$) for the men than for the women ($\hat{\sigma} = 6.7$), as shown in Table 7.2. Accounting for ranking in the model makes part of the heterogeneity observable, so that the remaining unobserved part σ is reduced. This applies to both men and women. But because β_- is much larger for the women than for the men, the reduction in σ is larger for the women, and this more than offsets the initial excess heterogeneity for the women.

Including quality in our model makes the deterministic part more complete and the random part less important. We see this reflected in the results, because both σ and ρ become smaller (closer to zero) when quality is included. The model explains more and is therefore better than the model of the previous chapter.

Other service characteristics

We perform the same analysis on other service characteristics, and the results are presented in Table 7.3. The table confirms the conclusions from Table 7.2: the estimated β_0 is not much affected when quality is included; ranking difference matters more than ranking sum; and heterogeneity is reduced by including quality.

The more complete analysis does, however, change the conclusion regarding hypothesis 6, which states that the probability that the service is in is the same in the men's singles as in the women's singles. The estimated probability of first service in for the average match is $\hat{\beta}_0 = 59.6\%$ with a standard error of 0.3% for the men, and 61.6% (0.5%) for the women. The difference is 2.0 (0.6), which is significant, while the earlier analysis (page 92) based on the model without rankings led to an insignificant difference. The reason is that the estimate for the women has slightly changed and that the standard errors have become smaller, because the inclusion of rankings has lowered the unexplained part. For the second service the probabilities for the average men and women remain remarkably

	β_0 mean	β_- $r_i - r_j$	β_+ $r_i + r_j$	σ heterog.	ρ correl.
Men					
1st service (s.) in	59.6	-0.1°	0.8°	5.0^*	-0.03°
2nd service in	86.5	1.5^*	0.7°	3.3°	0.11°
Points won if 1st s. in	74.0	8.1^*	4.0^*	5.9^*	-0.08°
Points won if 2nd s. in	59.6	8.2^*	1.3°	5.3^*	-0.41°
Points won on 1st s.	44.1	4.7^*	2.9^*	4.9^*	-0.36^*
Points won on 2nd s.	51.6	8.0^*	1.5°	5.2^*	-0.54^*
Women					
1st service in	61.6	0.3°	4.3^*	6.8^*	0.03°
2nd service in	86.7	4.6^*	4.5^*	4.8^*	-0.16°
Points won if 1st s. in	62.9	16.8^*	1.4°	5.0^*	-0.51°
Points won if 2nd s. in	52.8	13.5^*	-2.4°	7.7^*	-0.49°
Points won on 1st s.	38.7	10.6^*	3.7^*	3.6^*	-0.88^*
Points won on 2nd s.	45.8	14.1^*	0.4°	7.2^*	-0.48°

Table 7.3: *Service probabilities explained*

close. Our final conclusion is that hypothesis 6 is rejected for the first service, but not for the second.

The separation of p_i into its building blocks provides further insights. We first examine the β_- estimates. The reason (or at least one reason) why better players win more points is not that they hit more first services in (the estimates -0.1 and 0.3 are insignificant), but that they hit more second services in (1.5 and 4.6 are significant). Maybe players with a better ranking than their opponents go for a safer second service, assuming that they will win the resulting rally anyway. We cannot test this directly, because we do not observe what would have happened if a player had used his or her usual second service. However, our claim is somewhat supported by comparing the probability of points won if the second service is in to the probability of points won if the first service is in. The latter probability is 8.1 (men) and 16.8 (women) higher for the better server. The increments for the second service (8.2 and 13.5) are not significantly lower. Better players can therefore afford to play a safer second service. This does not mean that it is optimal to do so, because we do not know what would have been the winning probability if they had used their usual second service. Although better players hit more second services in, the more important reason why better players win more points is that they win more points *if* the service is in — a gain that does not come at the expense of more service faults. This is what makes top players top.

Next we examine the β_+ estimates. We consider matches where the players have the same ranking, $r_i = r_j$, while some matches are high-level ($r_i + r_j$ is large) and some are low-level ($r_i + r_j$ is small). Table 7.3 shows that in high-level matches the server wins particularly many points on first service, both for men and for women, because both 2.9 and 3.7 are significantly positive. The added detail reveals a remarkable difference between the men's singles and the women's singles. In the men's singles, the 2.9 increase is primarily determined by the increase in points won given that the service is in (4.0 is significantly positive), while the contribution from services in (0.8) is insignificant. On the other hand, in the women's singles, the 3.7 increase in primarily determined by the increase in services in (4.3 is significantly positive), while the increase in points won given that the service is in (1.4) is insignificant. We conclude that the cause for winning more points on first service differs between men and women: for the men, the main cause is that if the service is in, the point is more likely to be won; for the women, the main cause is that more services are in.

Aces and double faults

Aces and double faults differ from other service characteristics because they involve only the service and not the play following the service. The dependence on r_i is much stronger than on r_j, and it makes therefore more sense to write

$$\beta_i = \beta_0 + \beta_S r_i + \beta_R r_j$$

instead of

$$\beta_i = \beta_0 + \beta_-(r_i - r_j) + \beta_+(r_i + r_j).$$

This change of ordering does not affect the estimates of β_0, σ, and ρ, and the original β's can be recovered from

$$\beta_- = \frac{\beta_S - \beta_R}{2}, \qquad \beta_+ = \frac{\beta_S + \beta_R}{2}.$$

The sign of $\hat{\beta}_S$ in Table 7.4 shows, not unexpectedly, that better players serve more aces (especially in the men's singles) and fewer double faults (especially in the women's singles). The sign of $\hat{\beta}_R$ shows that the better is the receiver, the fewer aces are served in the women's singles. However, there is no evidence that the quality of the receiver influences the number of double faults (for either men or women) or the number of aces for the men.

	β_0 mean	β_S r_i	β_R r_j	σ heterog.	ρ correl.
Men					
Aces	8.8	6.8*	0.8°	5.0*	−0.16*
Double faults	5.5	−1.0*	0.2°	1.8*	−0.08°
Women					
Aces	3.2	3.0*	−3.1*	1.8*	−0.66*
Double faults	5.1	−4.0*	−0.5°	2.3*	−0.03°

Table 7.4: *Ace and double fault probabilities explained*

We return briefly to hypothesis 7 by testing whether the percentage of double faults is the same for men and women. The estimates for the average player are 5.45 (0.15) for the men and 5.10 (0.22) for the women. The difference is 0.35 and has a standard error of 0.27, leading to a confidence interval (−0.18, 0.88). The interval covers zero, just as the intervals on pages 80 and 92 where we tested the same hypothesis using simpler tools. The hypothesis is therefore not rejected, at least for the average player.

How do we reconcile this with the results from the previous section, in particular the fact that the probability of first service in is not the same for men and women, but the probability of second service in may be the same? The answer is estimation uncertainty. If one combines two differences, one of which is significant and one which isn't, then it may well happen that the difference of the combination (double fault) has a large standard error, so that it becomes insignificant. Our final conclusion regarding hypothesis 7 is therefore that we do not reject it (on average). It remains remarkable that the double fault probabilities are so close between men and women, given the differences between men's and women's tennis.

Further reading

Until 1967 only amateurs could compete at grand slam tournaments, but from 1968 onwards professionals could compete as well, and the 'open' era in tennis was born. In 1972 the leading professionals created the ATP, and one of their first acts was the establishment of a ranking system. The ATP Rankings began on August 23, 1973 and has continued to be the official ranking system in men's professional tennis until today. In 1973 Billie Jean King founded the WTA. The WTA Rankings started on November 3, 1975. Both ranking systems are published weekly. For a comparison of the ATP and WTA ranking systems to systems used in other sports, see Stefani (1997).

Although these two systems dominate tennis rankings, alternative systems have been proposed. Blackman and Casey (1980), for example, introduced a rating method for all serious tennis players — professionals and amateurs — that can be applied to define a tennis handicap, so that players with different skills can compete on even terms, just as in golf. Our book only uses the ATP and WTA rankings. We do not, however, use the rankings directly, but transform them using the expected-round concept (page 110), introduced in Klaassen and Magnus (2001).

The literature has tested whether variables other than rankings are relevant to tennis-winning probabilities. The overall conclusion is that the ranking difference is by far the most important determinant, and this conclusion is confirmed by our analysis. We briefly review this literature. Del Corral and Prieto-Rodríguez (2010) study the determinants of match wins at the grand slam tournaments 2005–2008. The most important determinant is the ranking, more precisely the difference in logarithmic rankings (following our expected-round concept). They also include other possible determinants, such as age, height difference, round, and right- and left-handedness. These do not appear to have much impact, except age: for the higher-ranked player the probability of winning the match is lower if he or she is older than the opponent.

The fact that Del Corral and Prieto-Rodríguez do not find a left-handedness effect, is somewhat contrary to common opinion, which is that 'lefties' have an advantage. Holtzen (2000) reports that about 10% of the general population is left-handed, and of professional tennis players also about 10%. But in the group of top players lefties are seriously overrepresented. Rod Laver, Jimmy Connors, John McEnroe, Rafael Nadal, Martina Navratilova, and Monica Seles are all left-handed, and over the period 1968–1999 left-handed players held the number-one ranking position about 35% of the time, both for men and for women. Roger Federer offered the explanation that left-handers typically get breakpoints on their favorite side, so that their swinging serves provide an advantage, especially when the receiver has a one-handed backhand. In terms of Chapter 4, we know that the average importance of points served in the ad court is higher, and this is precisely the side where lefties can serve out wide.

If left-handedness provides an advantage, this does not imply that in explaining or predicting point- or match-winning probabilities we need to take left-handedness into account *if we include the rankings in our analysis.* Any advantage for left-handed players will occur in every match, so that it will show up in the rankings. Therefore, if we correct for ranking, left-handedness does not matter much any longer. The suitability of this indirect way of dealing with left-handedness is confirmed by the results of Del Corral and Prieto-Rodríguez (2010). In addition, our unobserved quality correction π accounts for any remaining effect of left-handedness, as it does for any variable that is constant throughout the match.

A final potential determinant of quality is home advantage. Although less important than in team sports (such as football), home advantage may also exist in individual sports (such as tennis). Tennis players may be affected by crowd support and familiarity with local circumstances. Holder and Nevill (1997) find little evidence, but Koning (2011) finds an advantage, significant for men, insignificant for women. Whatever the truth, our unobserved quality correction π properly accounts for home advantage, because it remains constant throughout the match.

The test of hypothesis 12 on the competitiveness of men's versus women's tennis is used in Klaassen and Magnus (2001). Koning (2009) applies the test to more recent data (through 2008) and shows that competition in men's tennis is still tougher than in women's tennis. The results by Del Corral (2009) over 1994–2008 confirm this. These are *relative* results: men compared to women. Rohm *et al.* (2004) show that competitiveness in the men's singles at Wimbledon 1968–2001 has been high also in an *absolute* sense.

8
First and second service

Two services are allowed in tennis. This is an unusual feature compared to other sports with a service (table tennis, badminton, volleyball), and it allows intriguing questions — for example, whether a player is as good as his or her second service, as is often heard. Also, having two services offers a challenge for tennis players and investigators to think about the optimal balance between the two services. From a data point of view, it doubles the amount of information per player, allowing more precise analyses.

The optimal balance between the two serves requires a theoretical framework, which will be introduced in the next chapter. The current chapter offers a deeper analysis of the first and second service winning probabilities, and explains why we should move from an analysis of each service characteristic individually to a framework in which we examine the two characteristics jointly.

Is the second service more important than the first?

For some of the hypotheses considered so far it sufficed to use overall averages only. One hypothesis where this cannot be done is the following.

Hypothesis 13: *A player is as good as his or her second service.*

Overall averages are not useful here, because quality matters, so that we have to differentiate between players. We have to define when a player is 'good' and also when a second service is 'good'.

A service is 'good', as argued on page 75, when the percentage of points won on that service (not the percentage of services in or the percentage of points won if the service is in) is high. When a player is 'good' is less easy. We offer three interpretations.

Absolute quality

One possibility is to say that a player is 'good' if he or she is seeded. A player is then good compared to the group of players in the tournament as a whole without considering the opponent in the current match. For example, Serena Williams is good, irrespective of whether her opponent is Maria Sharapova or the world number 100. We call this *absolute* quality.

When a seed serves in the men's singles, taking seeded and non-seeded receivers together, the probability of winning a point on first service is 46.3%, while the probability on second service is 55.1%, a difference of 8.8%-points. The estimates are simply the relative frequencies in the data. When a non-seed serves, the percentages are 42.7% and 50.2%, a difference of 7.5%-points. Apparently, a seed in the men's singles wins particularly many points on his second service, in line with the hypothesis. But for the women the differences are 8.4% and 8.9% for a seed and non-seed, respectively. Hence, contrary to the hypothesis, a seed in the women's singles does not win especially many points on her second service.

Absolute quality corrected for quality difference

The previous interpretation distinguishes between the quality of servers, but not of receivers. This has the consequence that if we consider a non-seeded instead of a seeded server, then two things happen: we lower the quality of the server and also the quality difference between server and receiver. Perhaps these two changes have opposite effects, and combining them causes the above ambiguous result for men and women.

To find out, we now distinguish also between seeded and non-seeded receivers. Table 8.1 is the same as Table 7.1, except that we have separated the first and second service. We first compare Sd-Sd and NSd-NSd matches. With some simplification one may think of the players in such matches as being of the same strength: either both seeded or both non-seeded. So when comparing seeds to non-seeds, the quality difference is not affected. This is the

	Sd-Sd	Sd-NSd	NSd-Sd	NSd-NSd
Men				
Points won on 1st service	45.6	46.5	41.7	43.0
Points won on 2nd service	51.8	56.3	47.6	51.0
Women				
Points won on 1st service	41.0	42.8	34.1	37.1
Points won on 2nd service	46.0	52.2	40.4	47.0

Table 8.1: *Percentage of points won on first and second service, seeded (Sd) and non-seeded (NSd) players*

absolute point of view, but now corrected for the impact of quality differences.

In the men's singles, seeded players win many more points on their first service than non-seeded players (45.6% > 43.0%). The fact that seeds win more points is not as obvious as it may seem at first, because not only the service but also the opponent's return of service will be better. For the second service, the probability of winning a point differs much less (51.8% versus 51.0%). The situation is similar in the women's singles. There is a 3.9%-point difference between seeded and non-seeded players on the first service, and only −1.0%-point on the second service. It seems therefore that hypothesis 13 is not supported by the data. If anything, the hypothesis 'a player is as good as his or her *first* service' would be more appropriate.

Relative quality

In the previous analysis we compared Sd-Sd to NSd-NSd, so that there is variation in the overall quality of the match, but the quality difference is kept constant. We can also keep the overall quality constant and vary the quality difference. This is achieved by comparing Sd-NSd to NSd-Sd: the *relative* interpretation. In the men's singles, Table 8.1 shows that a seeded player wins 46.5% of points on his first service against a non-seed, while a non-seeded player wins 41.7% of his points on first service against a seed, a difference of 4.8%-points. On the second service the difference is larger: 56.3 − 47.6 = 8.7%-points. In the women's singles, the difference on the second service is also larger than on the first service (11.8 > 8.7). These comparisons suggest that the performance on the second service tells us more about a player's quality than the first.

We now have three interpretations and mixed outcomes. All three interpretations make sense. So is hypothesis 13 false or not?

Differences in service probabilities explained

To make further progress we reconsider Table 7.3, based on the model of pages 115 and 116.

	β_0 mean	β_- $r_i - r_j$	β_+ $r_i + r_j$
Men			
Points won on 1st service	44.1	4.7*	2.9*
Points won on 2nd service	51.6	8.0*	1.5°
Difference		−3.3*	1.4°
Women			
Points won on 1st service	38.7	10.6*	3.7*
Points won on 2nd service	45.8	14.1*	0.4°
Difference		−3.5°	3.3°

Table 8.2: *Difference between first and second service winning probabilities explained*

Table 8.2 contains the numbers from this table that are relevant to our investigation. The impact of r_i on each of the two service-winning probabilities depends on two variables: the quality difference ($r_i - r_j$) and the quality sum ($r_i + r_j$). Adding β_- and β_+ thus provides the impact of r_i when r_j is kept constant. This is the absolute effect. If we leave out the impact via $r_i - r_j$, we obtain the absolute effect corrected for quality difference, represented by β_+. Finally, β_- captures the relative effect. We discuss the results of all three interpretations in turn.

Regarding the absolute interpretation, we compute the sums of β_- and β_+. All four sums are positive, as expected, because higher-ranked servers (that is, servers with a higher r_i-value) score more points on first and on second service. The question is on which service they are particularly successful. For the men the difference between the first and second service sums is 1.9, and for the women 0.2, both in favor of the second service. These differences are, however, not significant, so there is no clear evidence in favor of hypothesis 13.

The second interpretation is obtained by filtering out the quality difference impact, thus focussing on the estimates of β_+. In the men's singles we find $2.9 > 1.5$, so that the higher the quality of the match the more points are won on first service compared to second service. The performance on first service tells us more about quality than the second service, contradicting hypothesis 13. The women's singles lead to the same conclusion: $3.7 > 0.4$. This is the same outcome as in the analysis based on seeds and non-seeds, but we have not yet examined the significance of these results. In the earlier analysis, based on seeds and non-seeds, the seeds perform significantly better on first service, while the performances on second service do not differ significantly. This is true for both men and women. In the current model-based analysis, the outperformance on first service is 1.4 for the men and 3.3 for the women, neither of which is significant. The reason why we get different results is that the standard errors computed in the first analysis are too small (see also page 91). Again we see the importance of proper statistical modeling for inference. We conclude that there is no compelling evidence against hypothesis 13 from the (corrected) absolute point of view.

Finally, we consider the relative approach by looking at β_-, which reflects the impact of the quality difference $r_i - r_j$. Table 8.2 shows that $4.7 < 8.0$ in the men's singles, so that the better the server relative to the receiver, the more points he wins on the second compared to first service. The second service performance now says more about his quality than the first service performance, in line with hypothesis 13. This is confirmed in the women's singles, and it is the same result as in the seeded versus non-seeded analysis. The difference of -3.3 for the men is significant, but the difference -3.5 for the women is not.

What have we learned from these three interpretations? That there is no compelling evidence against hypothesis 13, nor in the absolute sense (corrected or not), nor in the relative sense. In fact, there is some evidence in favor of the hypothesis in the relative interpretation. Hence we conclude that a player is as good as his or her second service, as the hypothesis states.

If it is true that players distinguish themselves more with the second service than with the first, then this has consequences for the possible change-of-rule discussed on page 28, namely to allow only one service. If only one service were allowed rather than two, then not only would the probability of winning a point on service decrease for all players, but also the difference between good and less good players would become larger. This is so, because the top players apparently distinguish themselves by their second service, and this is precisely the service that would remain after the rule change.

Joint analysis: bivariate GMM

One may object that our method of computing the standard error of the difference between the first- and second-service estimators is not entirely correct. In general, the variance of the difference between two estimators, say $\hat{\theta}_1$ and $\hat{\theta}_2$, is equal to the sum of the individual variances minus twice the covariance between the two estimators:

$$\operatorname{var}(\hat{\theta}_1 - \hat{\theta}_2) = \operatorname{var}(\hat{\theta}_1) + \operatorname{var}(\hat{\theta}_2) - 2\operatorname{cov}(\hat{\theta}_1, \hat{\theta}_2).$$

What we have done in the previous calculations is set the covariance equal to zero. However, the covariance will not be zero in general, because the first and second services are correlated — the better player will not only score more points on the first service, but also on the second service. The estimates themselves are not incorrect, but if we wish to compare them, then we need an estimate of the covariance. How can we generalize our 'complete' model, presented on pages 115 and 116, to include the covariance?

One source of correlation between the probabilities of winning a point on first and second service is related to the world rankings of the players. If the server has a top ranking, then both the probability of winning a point on first and on second service will be high, causing a correlation. Because we include the same ranking variables $r_i - r_j$ and $r_i + r_j$ in the models of both probabilities, this part of the correlation is observed and is accounted for in the estimation results so far.

The other source of correlation comes from π_i, the part of the probability that we do not observe. The form of the day, for example, affects the winning probabilities on both services, leading to correlation between the two probabilities. This correlation is not accounted for when the equations are estimated for each service probability separately, as we have done so far.

To properly account for the second source of correlation we combine the models for the first and second service-winning probabilities, allowing the two π_i-variables to be correlated. In other words, what we try to explain is no longer two separate probabilities, but rather a pair of two. Instead of our one-dimensional (univariate) GMM procedure we need a two-dimensional (bivariate) procedure — a little more complicated. Joint estimation (estimation of the bivariate model) automatically yields an estimate of the covariance between the estimators of the two β-parameters. An additional benefit is that joint estimation takes full advantage of the information that the data on the first service provide on the second service parameters and vice versa. This generally leads to smaller standard errors.

The estimates resulting from joint estimation are the same as the estimates in Table 8.2. This is not surprising because the moment conditions are the same. The additional estimates concern the correlation between the two π_i-variables. In the men's singles we find a correlation between first and second service probability of 0.52; in the women's singles it is 0.27, both positive, in line with our expectation.

Our main interest, however, is in another correlation, namely between the first- and second-service estimators of β_- and β_+. For β_- the correlation is 0.22 for the men and -0.07 for the women. For β_+ the correlations are 0.14 and 0.20, respectively. All four correlations are small (close to zero), which suggests that setting the covariance equal to zero, as we did in the previous section, may be the right thing to do. All results in Table 8.2 remain valid, and our conclusion at the end of the previous section, that a player is as good as his or her second service, remains valid as well.

Four service dimensions

In Table 8.1 we presented a breakdown of the probability p of winning a point on service in two components, one for the first and one for the second service. The probability of winning a point on the first service is itself made up of two components: the probability that the first service is in and the probability of winning the point if the first service is in. The same is true for the second service. In Table 8.3 we present this further breakdown, resulting in four dimensions.

	Sd-Sd	Sd-NSd	NSd-Sd	NSd-NSd
Men				
1st service in	58.7	59.5	59.4	59.5
2nd service in	87.8	87.3	85.6	86.2
Points won if 1st service in	77.7	78.1	70.2	72.4
Points won if 2nd service in	59.0	64.5	55.6	59.2
Women				
1st service in	65.6	61.5	60.5	60.2
2nd service in	88.8	89.1	85.2	85.0
Points won if 1st service in	62.5	69.6	56.4	61.7
Points won if 2nd service in	51.8	58.6	47.4	55.2

Table 8.3: *Four service dimensions, seeded (Sd) and non-seeded (NSd) players*

The percentages are in line with our earlier conclusions. Because of the problems that the corresponding standard errors may cause for inference, we immediately move to our model-based approach.

Four-variate GMM

One collection of model-based estimates of the four probabilities was presented in Table 7.3. These estimates were obtained for each probability separately. We know from the bivariate GMM section that estimating probabilities *jointly* improves the estimation results by exploiting the correlations between the π_i-variables. The four service characteristics considered here will, in general, be correlated. In particular, the correlation between the probabilities of winning the point if the first or second service is in will be positive, because being in a good form (that is, performing better than the ranking suggests, $\pi_i > 0$) will increase both winning probabilities. The same applies to the service-in probabilities: if a player hits more first services in, then very likely he or she will also hit more second services in.

We exploit information from the data on such correlations by estimating the four probabilities jointly, which results in a four-dimensional (four-variate) GMM procedure. This will be the most advanced model in this book. We shall need its estimates in the next chapter, but we can already use the model here to our advantage.

	β_0 mean	β_- $r_i - r_j$	β_+ $r_i + r_j$
Men			
1st service in	59.5	0.1°	0.6°
2nd service in	86.4	1.6*	0.7°
Points won if 1st service in	74.0	8.4*	3.4*
Points won if 2nd service in	59.4	7.9*	0.9°
Women			
1st service in	61.6	0.6°	3.9*
2nd service in	86.4	4.4*	4.3*
Points won if 1st service in	63.1	17.0*	1.2°
Points won if 2nd service in	52.6	13.8*	−2.2°

Table 8.4: *Four service dimensions explained in a four-variate analysis*

Table 8.4 presents a subset of the estimation results, namely the estimated value and significance of β_- (the impact of the quality difference $r_i - r_j$) and the estimated value and significance of β_+ (the impact of the quality sum $r_i + r_j$). The estimates of the correlations are in line with what we expect, and we do not present them.

The estimates in Table 8.4 are almost the same as those obtained from estimating the four characteristics one by one, as reported in

Table 7.3. This is good news and corroborates our findings based on the bivariate model. The conclusions drawn in earlier chapters therefore still apply. For example, as concluded from Table 7.3, if $\mathcal{I}$ is the better player in a match ($r_i > r_j$), then he or she wins more points not because $\mathcal{I}$ hits more first services in, a little because $\mathcal{I}$ hits more second services in, but mostly because $\mathcal{I}$ performs well if the service (first or second) is in.

Further reading

Hypothesis 13 has been analyzed earlier in Magnus and Klaassen (1999b).

9 Service strategy

Both players in a tennis match attempt to maximize the probability of winning the match. If points are independent, then each server chooses the service strategy that maximizes the probability of winning a point. A good strategy involves making both services neither too easy (in which case the receiver will kill it) nor too difficult (in which case the service will too often be a fault). We develop a model to answer the question of how difficult a player should make his or her service in order to maximize the probability of winning a point on service. We then test how close tennis professionals are to this maximum, that is, how efficient their service strategy is. We also ask whether top players are more efficient servers than other, lower-ranked, tennis professionals.

The server's trade-off

As always, we consider a match between $\mathcal{I}$ and $\mathcal{J}$. Player $\mathcal{I}$ serves and has to decide what type of first service to use and, if the first service is a fault, what type of second service. If $\mathcal{I}$ decides to go for a flat fast first service down the middle, and if the service happens to be in, then the probability of winning the point is high. But the probability that this service is in is low. If, on the other hand, $\mathcal{I}$ decides to go for an easy service, then the probability that the service is in is high, but the probability of winning the point if the service is in is low.

Hence there is a trade-off, and it is not clear what the overall effect of going for the easier service will be. This is already difficult if $\mathcal{I}$ had only one service at his or her disposal, and it becomes even more difficult with two services. The key probabilities in the trade-off are the probabilities introduced in the previous chapter. We denote them by

x_1: probability that first service is in,
x_2: probability that second service is in,
y_1: probability of winning the point if first service is in,
y_2: probability of winning the point if second service is in.

The probability p of winning a service point can then be computed as

$$p = x_1y_1 + (1 - x_1)x_2y_2,$$

in accordance with the discussion on page 75. Here we see the formalization of the trade-off. Increasing x_2 has the direct effect of increasing p, and also the indirect effect of lowering y_2 and hence decreasing p. This is the trade-off for the second service. The same applies to the first service, but here there is an additional effect, because increasing x_1 leads to a lower value of $1 - x_1$, thus reducing the possible contribution of the second service to winning the point. The question in this chapter is how to determine the player's optimal service strategy (x_1, x_2).

The question can easily become too complicated to answer. Thus we impose restrictions to keep the problem manageable, in accordance with Albert Einstein's words: simple but not too simple. One assumption that we already imposed is that players serve to maximize the probability of winning. This seems a reasonable assumption, especially for professional players. Amateur players, however, may have a slightly different objective. Casual observation suggests that amateurs often fully attack the first serve, making the service more difficult than their skills allow, so that the first service is frequently out or in the net. But the satisfaction from serving the odd ace makes up for this inefficient service behavior. Amateurs typically also make their second service too easy, so as to avoid the disappointment of a double fault. For amateurs, satisfaction may be more important than winning. It is unlikely, however, that professionals behave this way. They want to win.

Another restriction is that we confine the analysis to the four key probabilities x_1, x_2, y_1, and y_2. We realize that, for both services, x and y depend on several variables, such as speed, direction, spin, concentration, emotions, and so on. Since we have no data on these additional variables, we assume that the variables that matter for x are chosen optimally given the value of x. Since we study the world's best tennis players this assumption seems reasonable, and it allows us to act as if only x matters for y and hence for p, so that the server has to choose only x_1 and x_2.

The y-curve

We shall develop a mathematical model for the server. A key element in this model is a function $y(x)$ relating the probability x that the service is in to the probability y that $\mathcal{I}$ wins the point if the service is in. There are two services but there is only one y-curve (per player, because each player may have a different curve). The two services represent different points on that curve: $y_1 = y(x_1)$ and $y_2 = y(x_2)$ for the first and second services, respectively.

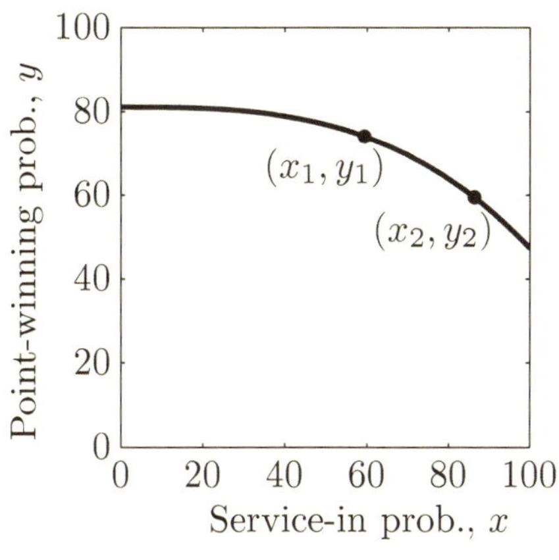

Figure 9.1: *The y-curve*

To illustrate, consider the estimated probabilities for an average match in the men's singles, as reported in Table 8.4. We find

$$(x_1, y_1) = (0.595,\ 0.740), \qquad (x_2, y_2) = (0.864,\ 0.594).$$

Figure 9.1 shows how the y-curve might look for the men with dots indicating the two (x, y) combinations. It seems reasonable that the easier a player makes his or her service, the more likely it is that the service is in (x increases), but the less likely that the point is won if the service is in (y decreases). Hence, we expect y to be a decreasing function of x.

For the average women's match the two points on the y-curve are

$$(x_1, y_1) = (0.616, 0.631), \qquad (x_2, y_2) = (0.864, 0.526).$$

The y-curve is lower for the women than for the men. Also, since the slope $(y_2 - y_1)/(x_2 - x_1)$ of the line connecting both points equals -0.54 for the men and -0.42 for the women, the y-curve is less steep for the women than for the men. This is consistent with the idea that for women the service is less influential, so that varying the service difficulty (x) has a smaller impact (y) on winning the point.

Optimal strategy: one service

To explain the idea of finding the optimal service strategy, we first consider the hypothetical case of one service, where the player has to determine only one optimal x. Of course, we do not know the y-curve for a specific match. But suppose we did. Then we would argue as follows. First, define

$$w(x) = x \cdot y(x),$$

which transforms the conditional probability y (of winning the point given that the service is in) into the unconditional probability w (of winning a point on that service).

Figure 9.2 illustrates the w-function. At $x = 0$ we have $w(0) = 0$. When x increases, y decreases, but the server now has at least some positive probability of winning the point, so w increases. When x increases further, y continues to decrease. This is the trade-off described earlier. When x reaches the point (indicated in the figure by x_2^*) where the positive impact of the higher x on w is exactly offset by the negative indirect effect via the lower y, then w reaches its maximum. When x increases beyond that point, then the decrease in y dominates and w decreases until $x = 1$. The function $w(x)$ looks like a parabola with a top.

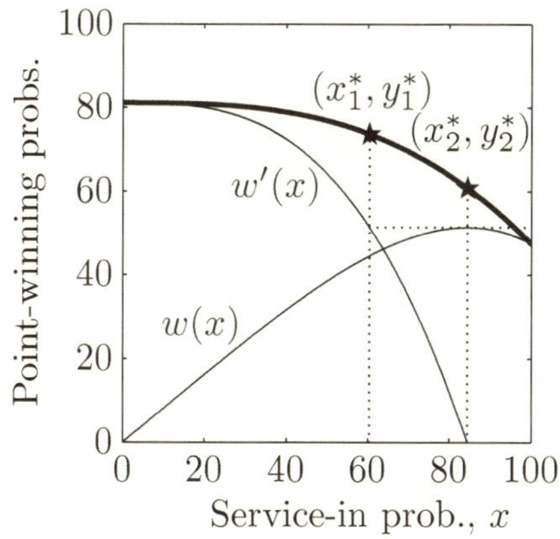

Figure 9.2: *The y-curve and the optimal service strategy*

In the single-service case, $w(x)$ represents the probability of winning the point. The optimal service strategy is obtained by maximizing $w(x)$ with respect to x. This is achieved by differentiation, and we find the optimum as the solution of

$$w'(x) = 0.$$

This is the value of x, denoted x_2^* in the figure, where the two elements in the trade-off exactly offset each other: $y(x) = -xy'(x)$.

Figure 9.2 also contains the derivative function $w'(x)$. This curve crosses the horizontal axis precisely below the top of the parabola, because $w'(x) = 0$ occurs at $x = x_2^*$ where the probability of winning the point is maximal. One may wonder what happens if the top of the parabola lies to the right of $x = 1$, so that $w'(x) = 0$ has no solution in the interval $(0, 1)$. In that case, $w(x)$ has a maximum at $x = 1$, and the player should try to hit all services in.

Optimal strategy: two services

The problem is more complicated with two services, although the idea remains the same. Based on the definition of $w(x) = xy(x)$, the probability of winning a point on the first service is $w_1 = w(x_1)$, and the probability of winning a point on the second service is $w_2 = w(x_2)$. As a result, the probability that $\mathcal{I}$ wins the point is given by

$$p(x_1, x_2) = w(x_1) + (1 - x_1)w(x_2).$$

The optimal service strategy, which we denote (x_1^*, x_2^*), is obtained by maximizing $p(x_1, x_2)$ with respect to x_1 and x_2. This is achieved by taking partial derivatives and setting them equal to zero. The two partial derivatives are found by first differentiating p with respect to x_1 while keeping x_2 constant,

$$w'(x_1) - w(x_2) = 0,$$

and then differentiating p with respect to x_2 while keeping x_1 constant,

$$(1 - x_1)w'(x_2) = 0.$$

The optimal service strategy (x_1^*, x_2^*) is thus given as the solution of the two equations

$$w'(x_1) = w(x_2), \qquad w'(x_2) = 0,$$

and this solution can be found in three steps. First, solve the equation $w'(x) = 0$ and call the solution x_2^*. Next calculate $w_2^* = w(x_2^*)$. Finally, solve $w'(x) = w_2^*$ and call the solution x_1^*.

The geometry of the solution is illustrated in Figure 9.2. The maximum of w occurs at $x = x_2^*$, where $w'(x) = 0$. This implies that x_2^* in the two-service problem is the same as the optimal single service in the hypothetical one-service problem, in line with our discussion on page 29. The tangent of the curve $w(x)$ at the maximum $x = x_2^*$ has a level $w(x_2^*)$, and the intersection of this horizontal line through $w(x_2^*)$ with the curve $w'(x)$ is at $x = x_1^*$. The optimal probabilities are therefore (x_1^*, y_1^*) and (x_2^*, y_2^*).

Existence and uniqueness

In deriving the optimal strategy (x_1^*, x_2^*) we have implicitly assumed that an optimal strategy exists and that there is only one. What restrictions on the y- and w-curves are required to ensure existence and uniqueness? The y-curve in Figure 9.2 is decreasing and concave. It is decreasing because a higher value of x is associated with a lower value of y, and it is concave because, if we connect any two points of the curve by a straight line, then this line segment lies below the curve. Concavity is typical for a y-curve, although it is not essential. What is essential is that w is concave. These are the two conditions that need to be imposed: y is decreasing and w is concave. Together, the two conditions imply the existence and uniqueness of a solution.

We discussed above why it is reasonable that y is decreasing. But what does it mean that w is concave? The concavity of w reflects the fact that if a player's service is too difficult, then he or she will lose the point because it is a fault, but if the service is too easy, then he or she will also lose the point because the receiver hits a return winner. More specifically, when x increases starting at $x = 0$, then $w(x)$ increases until some point $x = x_2^*$, and then decreases until $x = 1$. This is the essence of concavity and it implies that there exists an optimal (second) service, neither too easy nor too difficult, which maximizes the player's probability of winning the point on that service.

Four regularity conditions for the optimal strategy

Imposing these two conditions (y is decreasing and w is concave) has four implications for the optimal strategy (x_1^*, x_2^*):

(R1) $x_1^* < x_2^*$,
(R2) $y(x_1^*) > y(x_2^*)$,
(R3) $w(x_1^*) < w(x_2^*)$, and
(R4) $w(x_2^*) - w(x_1^*) < (x_2^* - x_1^*)w(x_2^*)$.

Conditions R1 and R2 mean that the first service should be more difficult than the second service in two ways: it is less often in, and if it is in, it is more likely to win the point. Regularity condition R3 says, as we shall show in a moment, that always using the first service (so using service type x_1^* for x_1 *and* x_2) is not optimal. Similarly, condition R4 means that always using the second service is not optimal.

Let us formally prove these statements. Since $w'(x_2^*) = 0$ and $w'(x_1^*) = w(x_2^*) > 0$, it follows that $w'(x_1^*) > w'(x_2^*)$. Now, w is concave and hence its derivative w' is a decreasing function. This implies that $x_1^* < x_2^*$ and hence that R1 holds. Condition R2 follows from R1 because y is a decreasing function.

Next, if (x_1^*, x_2^*) is the optimal strategy, then the strategy (x_1^*, x_1^*) (always use the first service) is less than optimal. In other words, $p(x_1^*, x_2^*) > p(x_1^*, x_1^*)$. Using the formula for $p(x_1, x_2)$ on page 142 we then find that

$$w(x_1^*) + (1 - x_1^*)w(x_2^*) > w(x_1^*) + (1 - x_1^*)w(x_1^*),$$

which is equivalent to R3. Similarly, the strategy (x_2^*, x_2^*) (always use the second service) is less then optimal, so that $p(x_1^*, x_2^*) > p(x_2^*, x_2^*)$. This leads to

$$w(x_1^*) + (1 - x_1^*)w(x_2^*) > w(x_2^*) + (1 - x_2^*)w(x_2^*),$$

which is equivalent to R4.

If a player serves optimally, he or she should satisfy the four regularity conditions. These conditions may seem fairly obvious characteristics of service behavior, but are they actually satisfied in practice?

	R1	R2	R3	R4	All four
Men	100	91	78	80	59
Women	98	72	77	64	42

Table 9.1: *Empirical realization (percentages) of the four service regularity conditions R1–R4*

We use the observed relative frequencies of each player in each match of our Wimbledon data set as a first step to find out how many players serve according to these four conditions. Table 9.1 reveals that the conditions are often *not* satisfied. The condition $x_1 < x_2$ appears to be almost always satisfied, which means that almost all players take more risks on their first service than on their second service (as they should). However, this additional risk does not necessarily translate into higher productivity: the condition $y(x_1) > y(x_2)$ is only satisfied for 91% of the men and 72% of the women. Condition R3 requires that (x_1, x_2) is a better service strategy than (x_1, x_1), but this is only true for 77 to 78% of the players. So, for 22 to 23% of the players hitting two first services would win more points than hitting the traditional first and second service. Condition R4 requires that (x_1, x_2) is a better service strategy than (x_2, x_2), but this is only true for 80% of the men and 64% of the women. For only 59% of the men and 42% of the women are all four consistency requirements satisfied. It therefore seems that for many players the probability of winning a point can be increased by changing their service strategy.

This conclusion is, however, too simplistic. Our theory is in terms of probabilities, whereas our observations are relative frequencies. Relative frequencies are not the same as probabilities. The difference is 'noise'. We know from the analysis on page 100 that noise plays a substantial role in summary statistics, and this is confirmed here. But noise can be modeled. The statistical theory of the method of moments developed in Chapter 6 allows us to do so, and later in this chapter we shall combine this statistical theory with the mathematical model we are currently developing.

Functional form of y-curve

So far we have put some restrictions on the y-curve, but we have not specified a functional form. It will be convenient to do so. The simplest specification is a linear function. This is simple, but too simple, because it leads to results that are not credible. For example, it forces $x_1^* \leq 1/2$, which is not realistic since the estimated x_1 is 59.5% for men and 61.6% for women.

Some curvature needs to be introduced. A linear specification requires two parameters, and the simplest extension therefore requires three parameters. A suitable candidate is

$$y(x) = \gamma_0 - \gamma_1 x^\lambda \qquad (\lambda > 0).$$

If the two gammas (γ_0 and γ_1) and the lambda (λ) are known, then the y-curve is completely specified and the optimal service strategy can be calculated.

Tennis allows two services and the data provide information on x_1, x_2, y_1, and y_2. The resulting points (x_1, y_1) and (x_2, y_2) must lie on the y-curve, and hence we can solve for two parameters. The fact that we have two services is a unique feature of tennis. If there were only one service, as is the case in many sports, then we could only solve for one parameter. Having two services doubles the amount of information on the y-curve.

For given λ, we can thus solve γ_0 and γ_1 from the two equations

$$y_1 = \gamma_0 - \gamma_1 x_1^\lambda, \qquad y_2 = \gamma_0 - \gamma_1 x_2^\lambda,$$

and this gives

$$\gamma_0 = \frac{y_1 x_2^\lambda - y_2 x_1^\lambda}{x_2^\lambda - x_1^\lambda}, \qquad \gamma_1 = \frac{y_1 - y_2}{x_2^\lambda - x_1^\lambda}.$$

Hence, in order to compute the y-curve and from there the optimal service strategy, we require two things. First, we need the probabilities (x_1, y_1) and (x_2, y_2), that is, the probabilities actually employed by the player. These determine two points on the y-curve. Second, we need λ to specify the curvature of the y-curve.

Efficiency defined

If we know x_1, x_2, y_1, and y_2, then we can compute the probability p that the server wins a point using the formula on page 138. If we also know the curvature parameter λ, then we can compute γ_0 and γ_1, so that we know the whole y-curve. Once we know the y-curve, we also know the optimal strategy (x_1^*, x_2^*) and the corresponding optimal y-values $y_1^* = y(x_1^*)$ and $y_2^* = y(x_2^*)$. From these optimal values we can compute the maximum probability p^*.

We define the service efficiency of a player in a given match as

$$p/p^*.$$

This is a number between zero and one, and the closer p/p^* is to one, the higher is the efficiency. Of course, the efficiency differs per player and depends also on the opponent.

Now that we have defined efficiency we can formulate our next hypothesis.

Hypothesis 14: *Players have an efficient service strategy.*

Naturally, for a player in a given match, the realized value of p will be lower than the optimal value p^*. But how much lower? Is p close to p^*, so that the difference is irrelevant? Or is the difference substantial? If so, are there differences between men and women, and between higher-ranked and lower-ranked players? To answer these questions and test the hypothesis, we need estimates of the key probabilities x_1, y_1, x_2, and y_2, and of the curvature parameter λ.

Efficiency of the average player

We have no estimate of λ but we do have, at least for the average match, quite precise estimates of the key probabilities from the GMM analysis in the previous chapter. These estimates could possibly provide a first impression of the efficiency of tennis players. In the average men's singles we have

$$(x_1, y_1) = (0.595,\ 0.740), \qquad (x_2, y_2) = (0.864,\ 0.594).$$

Fixing $\lambda = 3$ gives $\gamma_0 = 0.811$ and $\gamma_1 = 0.336$, so that the y-curve is now known. The optimal strategy becomes $(x_1^*, x_2^*) = (0.605,\ 0.845)$, which implies $y_1^* = 0.737$ and $y_2^* = 0.608$. This is the situation illustrated in Figure 9.2. The optimal probability of winning a point on service is $p^* = 0.649$.

This is for $\lambda = 3$. If we take another value for λ then the results change. For $\lambda = 2$ the optimal strategy becomes $(x_1^*, x_2^*) = (0.567,\ 0.884)$ and $p^* = 0.649$. For $\lambda = 4$ we find $(x_1^*, x_2^*) = (0.630,\ 0.825)$ and $p^* = 0.651$. The optimal strategy varies substantially with λ, but the optimal overall probability p^* of winning a point on service appears to be quite robust. This is confirmed by the results for the average women's match, where p^* is 0.565, 0.563, and 0.564 for λ equal to 2, 3, and 4, respectively. The conclusion is that we should be careful with conclusions regarding the optimal strategy unless we have a good estimate of λ, but that we can safely use the model to quantify efficiency, which is our current focus.

The efficiency of the average server, when $\lambda = 3$, is given by $p/p^* = 0.9996$ in the men's singles and 0.9998 in the women's singles. It is tempting to conclude that tennis professionals are almost perfectly efficient, and that hypothesis 14 is true. We should realize, however, that the value of p/p^* here refers to the efficiency of the average player. All players differ from the average player, and it is the individual players that we are interested in. What we need to do is compute individual efficiencies. Then, if we are interested in an overall result, we can aggregate the individual efficiencies. We want the average efficiency, not the efficiency of the average.

Observations for the key probabilities: Monte Carlo

The development so far provides us with a mathematical model, which can be used to compute efficiency if we know the four key probabilities (x_1, x_2, y_1, and y_2) and the curvature parameter λ. These parameters are not known and we need to estimate them, thus moving from a mathematical to a statistical model. Regarding the curvature parameter we shall assume that it is constant over players (but different for men and women), and we estimate the parameter as $\hat{\lambda} = 3.07$ for the men and $\hat{\lambda} = 3.83$ for the women. But the four key probabilities are not constant. They differ over individual players. We don't know these probabilities. All we have are the corresponding relative frequencies.

The GMM estimation theory developed in Chapter 6 allows us to estimate the mean and variance of the four key probabilities as a function of a player's ranking r_i and the opponent's ranking r_j. This led to Table 8.4. The estimates in that table do not depend on λ, although the underlying mathematical model does. A natural question is whether we can safely use these estimates. The answer is that we can. To understand why, consider Figure 9.1. The data provide information on the location of the points (x_1, y_1) and (x_2, y_2), and varying λ affects the curvature of the y-curve but not the points themselves. This is the reason why we can ignore λ when estimating the location of the points, and it is important, because it allows us to separate the GMM procedure from the determination of λ and the efficiency analysis, and it validates the use of Table 8.4.

Application of the mathematical model requires estimates of the key probabilities, but we have only estimated their mean and variance. Somehow we have to *generate* representative observations. This is achieved by a method called 'Monte Carlo', where we 'draw' from a normal distribution based on the estimated mean and variance. We let the computer generate numbers from the estimated distribution. These numbers, by construction, share the properties of the distribution. In particular, they have the same mean and variance, and can therefore be considered realistic values of the service probabilities x_1, x_2, y_1, and y_2. Each draw provides us with eight probabilities, namely four service probabilities for each of the two players in one match. Given the estimate for λ, we then have an estimate of the y-curve for both players, and hence we can calculate the optimal probabilities and efficiency for the two players in this match. We repeat the procedure for every match played at Wimbledon in the years covered by our data set.

This is a complicated procedure. What we have done is to replace each actual match by an artificial match, but one with the same properties as the actual match. To reduce the randomness involved in the procedure we don't rely on just one draw per match, but we draw fifty times. We could have drawn more times (we don't need more data to do this, only more computing time), but fifty times provides a sufficiently accurate coverage of the distribution.

Efficiency estimates

The Monte Carlo method provides us not only with an estimate of the actual probability p, but also of the theoretically optimal probability p^* and hence of the efficiency p/p^*. The average efficiency is 98.9% (with standard error 0.2%) for the men and 98.0% (0.3%) for the women. The service strategy of professional tennis players is therefore quite efficient, but not fully efficient, and hypothesis 14 should be rejected.

Not only do we reject the hypothesis, but we also quantify the inefficiency: 1.1% for men and 2.0% for women. Apparently, the women are less efficient than the men, at least in choosing their service strategy. The inefficiencies are small, but not that small if one investigates the impact on the paycheck (see page 152).

The estimated efficiencies 98.9% (men) and 98.0% (women) are averages. The average is, however, only one aspect of the distribution of efficiency across players. The Monte Carlo draws allow us to estimate the complete distribution in the form of their density (recall from page 93 that a density is a smoothed histogram), showing how often each particular level of efficiency p/p^* occurs.

Figure 9.3 presents a solid line and a band around it. The line is the estimated density and the band is the 95% confidence interval, reflecting the uncertainty of the estimation procedure. The tighter the band, the more accurate the density is estimated.

The density tells us that 25% of the men have an inefficiency of more than 1.4% and that 5% of the players have an inefficiency of more than 3.3%. For the women, 25% of the players have an

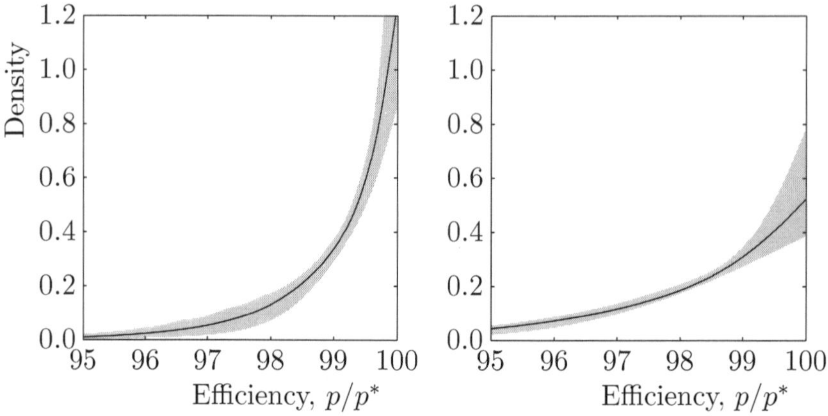

Figure 9.3: *Distribution of efficiency across players (men left, women right)*

inefficiency of more than 2.8% and 5% have an inefficiency of more than 5.8%. Inefficiencies thus vary substantially across players.

Mean match efficiency gains

The fact that hypothesis 14 is rejected does not imply that the inefficiencies are large. In fact, they appear to be small. It is difficult to define what exactly we mean by 'small'. But we can try and answer the question indirectly by studying the impact of inefficiency, and we now analyze this impact at point, game, and match level.

At point level, the impact of inefficiency is that by serving efficiently, men can increase p by 0.7%-points (from 64.8% to 65.5%) on average, and women by 1.2%-points (from 56.3% to 57.5%).

If we consider a game, then the impact of inefficiency becomes larger, not because the players perform differently but because of the structure of the tennis scoring system, which causes a magnification effect (see page 16). The gain of 0.7%-points at point level becomes 1.1%-points at game level for the men, and 1.2%-points at point level becomes 2.5%-points at game level for the women.

At match level (arguably the most natural unit) the impact of inefficiency does not only depend on the inefficiency of player $\mathcal{I}$ but also of player $\mathcal{J}$. What would the efficiency gain for player $\mathcal{I}$

be if he or she switches to serving efficiently while player $\mathcal{J}$ does not? The answer is that the mean increase in the probability of winning the match is 2.4%-points for the men and 3.2%-points for the women. This is not so small anymore. If, by serving efficiently, a player can change the match-winning probability from 50-50 to 53-47, then this represents a real gain.

Efficiency gains across matches

The mean match efficiency gain averages out differences in efficiency gains across matches. Each match is different and these differences are ignored by just considering the mean. In a well-balanced match, for example, serving efficiently is expected to be more important than in a match where one player is much stronger than the other. This issue is illustrated in Figure 9.4.

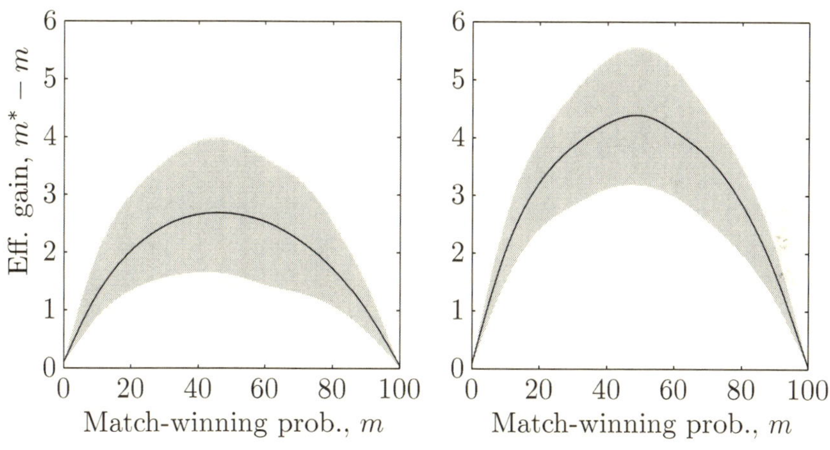

Figure 9.4: *Efficiency gain as a function of the match-winning probability (men left, women right)*

For a match between players $\mathcal{I}$ and $\mathcal{J}$, Chapter 2 explains how we can use point-winning probabilities p_i and p_j to compute the probability m_i (here simply denoted by m) that $\mathcal{I}$ wins the match. This is when both players serve normally. If player $\mathcal{I}$ serves optimally while $\mathcal{J}$ continues to serve normally, then the winning probability increases to m^*. Each value of m can be generated by many different combinations of p_i and p_j, and each combination yields its own value of m and m^*. For each m, there is therefore no unique value of $m^* - m$, but rather a whole distribution of values. The solid line in the figure presents the median of these values, at each value of m. The figure also contains the 95% confidence interval around the median, which again measures the impact of estimation uncertainty.

If two players are approximately of equal strength then the median efficiency gain at match level for the efficient server is 2.7%-points for the men and 4.4%-points for the women. This is the median gain. In some matches the gain of serving efficiently is much higher. For example, 10% of the players in a balanced match will have an increase of more than 10%-points in the men's singles (15%-points in the women's singles).

In very uneven matches, it does not matter whether the server increases the efficiency or not. The figure shows, at m close to zero or one, that the impact $m^* - m$ of serving efficiently is virtually zero: the top player wins anyway, even when his or her service is not fully efficient.

Impact on the paycheck

Players play one or more matches in a tournament until they lose. Then they get a paycheck. If by serving efficiently a player wins instead of loses the current match (and possibly also the next match), then the prize money may increase substantially, particularly at grand slam tournaments.

In order to study the monetary impact of serving efficiently, we do a little experiment. We run a hypothetical tournament of 128 players (seven rounds, like Wimbledon), where in each match both players have probability 50% to win the match, except one player who serves efficiently. We have just seen that for two players of equal strength the median efficiency gain for the efficient server is 2.7%-points for the men and 4.4%-points for the women. We therefore assume in our experiment that the only efficient player has 52.7% (that is, an additional 2.7%-points) probability of winning a match in the men's singles, and 54.4% in the women's singles. What is the expected monetary gain for the efficient player?

In grand slam tournaments the paycheck approximately doubles in each round. If we assume that this is exactly true, then the expected paycheck for the efficient player will rise by 18.7% for men and by 32.8% for women. At Wimbledon this would mean an expected additional income of approximately $10,000 for the efficient man and $15,000 for the efficient woman. Hence, even though the inefficiency at point level may seem small, the monetary benefit of efficiency can be substantial.

Why are players inefficient?

We conclude that the service strategy for professional tennis players in a top tournament is not efficient, and hence that hypothesis 14 should be rejected. The rejection of efficiency is not large but its (monetary) effects can be substantial.

Inefficiency is measured by the maximum possible relative increase in the probability of winning a point on service, and we estimated it to be 1.1% for men and 2.0% for women on average. The impact of serving efficiently can be quantified at various levels of aggregation. At point level the impact is 0.7%-points for men (1.2%-points for women), at game level 1.1%-points (2.5%-points for women), and at match level 2.4%-points (3.2%-points for women). These differences reflect the scoring system and the fact that at match level the impact of service efficiency depends on the quality difference between the players. In terms of expected monetary gains the expected paycheck for the efficient player could rise by 18.7% for men and 32.8% for women. So even small inefficiencies can have substantial financial consequences. This, in summary, is what we found. Higher-ranked players are also more efficient than lower-ranked players, and the closer the match, the more efficient a player serves.

But what is the reason for this inefficiency? Perhaps professional tennis players know their y-curve, but are not able to solve the optimization problem. Or they correctly solve the optimization problem, but on the wrong y-curve. From the point of view of achieving optimality it is much easier for a server to work out the optimal second service (maximize $w(x)$, as in the hypothetical case of a single service) than to work out the optimal first service, because the latter requires knowing the optimal second service. We will use this difference to gain further insight into the reasons for inefficiency.

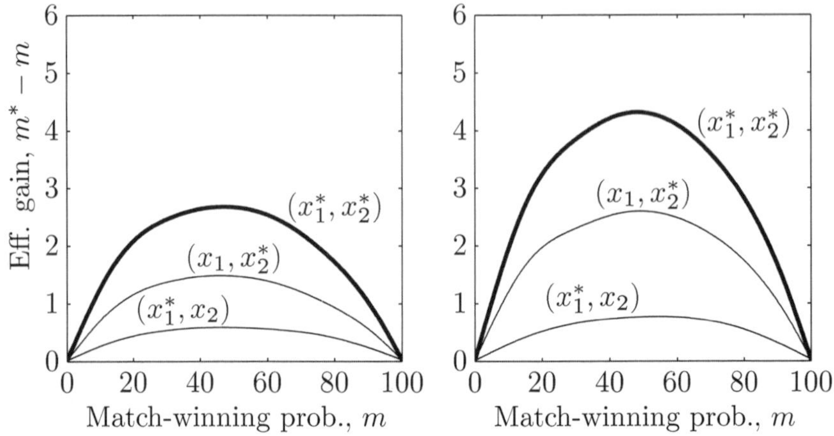

Figure 9.5: *Decomposition of efficiency gain into first and second service (men left, women right)*

Let us separate first-service and second-service efficiency, and ask which service is more efficient on average. In Figure 9.5 we plot three curves for the men (left) and the women (right). The top graph labeled (x_1^*, x_2^*) is the same as the median in Figure 9.4. We now decompose this graph into (x_1, x_2^*) where only the second service is optimal, and (x_1^*, x_2) where only the first service is optimal.

What do we see? Players can achieve a larger efficiency gain on the second service than on the first service, in spite of the fact that the second service is conceptually easier to optimize and should therefore be closer to its optimum. The fact that the second service is less efficient than the first service provides evidence that, although players may be maximizers, they do not maximize the correct function.

Rule changes

Rule changes can affect the nature of tennis and the skills required to win a match. The International Tennis Federation, in charge of the Rules, is eager to know the possible impact of rule change proposals. Can our mathematical model help to predict the effect of rule changes?

The optimal service strategy (x_1^*, x_2^*) follows from maximizing the probability of winning a point on service $p(x_1, x_2)$. Hence, if a rule change does not affect $p(x_1, x_2)$, then there will be no effect on the optimal strategy. If $p(x_1, x_2)$ is affected, then the optimal strategy may well change. To find out how, we need to redo our optimization calculations, described on page 142.

Let us reconsider the rule changes discussed in Chapter 2. Neither the match tiebreak of ten points, nor the short set until four games, nor the no-ad rule have any impact on $p(x_1, x_2)$, and hence the optimal strategy (x_1^*, x_2^*) is unaffected. This is not surprising, because these proposals only affect the number of points needed to win the match, not the way in which a player can win a point. Adjusting the strategy away from (x_1^*, x_2^*) in these cases would only lead to a *lower* value of p.

Things are different when the second service is abolished, because this rule change does affect the nature of tennis. The question is what will be the optimal probability x of hitting the single service in? In terms of our model, what will be the value of x that maximizes the probability $w(x)$ of winning a point when using a service of type x? As derived on page 142, and in line with the analysis on page 29, the best a player can do is to use the current second service as the only service in the new regime.

Serving in volleyball

Rule changes are rare in tennis, but less rare in many other sports. A comparison with volleyball is of particular interest, because in volleyball a rule change occurred that involved the service. Originally, if the serving team won the rally, the team would score a point, but if they lost the rally, the score would not change but the service would go to the other team. This is the 'side-out' scoring system. From 1999 onwards, after pressure to make volleyball faster-paced and more attractive to television, volleyball uses 'rally' scoring, in which teams gain a point whenever they win the rally. Many volleyball coaches reacted to this switch by telling their players to be more cautious when serving, because a service fault would immediately give a point to the opponent.

Is this coaching advice correct? Maybe our model can help in answering this question. Let x be the probability of hitting a service in, as before, but now for a volleyball team, and let $v(x)$ be the probability that the serving team wins the rally using a service of type x. In the rally-point system, team $\mathcal{I}$ can only win the point by winning the current rally, and this has probability v_i. In the old system (side-out), team $\mathcal{I}$ could win a point on service in several ways. First, by winning the current rally with probability v_i. Second, by losing the current rally, winning the next rally on the opponent's serve, and then winning the third rally on its own service. The probability of this event, assuming independence of points, is $(1-v_i)(1-v_j)v_i$, which we write as $v_{ij}v_i$ with $v_{ij} = (1-v_i)(1-v_j)$. Third, by losing the own serve, winning the opponent's serve, losing the own serve, winning the opponent's serve, and finally winning the own serve. The probability of this happening is $v_{ij}^2 v_i$. In fact, there are infinitely many possibilities. The sum of all corresponding probabilities is the probability of winning a point:

$$p_i = v_i + v_{ij}v_i + v_{ij}^2 v_i + \cdots = v_i \left(1 + v_{ij} + v_{ij}^2 + \cdots\right) = \frac{v_i}{1 - v_{ij}}.$$

Since v_i and v_j are probabilities, they are bounded by zero and one, and we may safely assume that they are not zero or one. Hence $0 < v_i < 1$ and $0 < v_j < 1$, implying that $0 < v_{ij} < 1$ and that p_i in the previous formula is well-defined.

This then is the probability of winning a point on service for team $\mathcal{I}$ in the side-out scoring system. In the rally-point system the probability is simply $p_i = v_i$. This shows that p_i is affected by the introduction of the rally-point system, and hence that the optimal service strategy x^* may change. Whether the optimal service in fact does change can be found out by the use of a little calculus.

The optimum x^* is the solution of $p_i'(x) = 0$. Under rally-point scoring this is the solution of $v_i'(x) = 0$. Determining the optimum under side-out scoring is more involved. Using the fact that v_j does not depend on x, we find

$$p_i'(x) = \frac{(1 - v_{ij})v_i'(x) + v_i v_{ij}'(x)}{(1 - v_{ij})^2} = \frac{v_j v_i'(x)}{(1 - v_{ij})^2}.$$

Since $v_j \neq 0$ and $v_{ij} \neq 1$, the only way that $p_i'(x)$ can be zero is that $v_i'(x)$ is zero. But this is the same condition as under rally-point scoring! Hence the solution x^* is the same as well. A little mathematics thus produces the remarkable result that the rule change affects the objective function $p(x)$, but not the optimal service strategy.

What happened to coaching? A year or so after the introduction of rally-point scoring, the coaches realized that being more cautious when serving was the wrong strategy, and they advised players to serve exactly as before. Their initial advice — use a safer service so a higher x — has the benefit of reducing service faults and direct point losses, but it has the cost that the team will lose the rally more often if the service is in. Our little model shows that at $x = x^*$ benefit and cost are balanced in both the side-out system and the rally-point system. A safer service is ill-advised and will lead to a worse performance.

Further reading

It is slightly amusing and not very realistic to believe that tennis players actually calculate partial derivatives and work out optimal strategy in the way our mathematical development suggests. In the context of billiards, Friedman (1953, p. 21) claims that an expert billiard player makes his shot as if he knew the complicated mathematical formulas that would give the optimum direction of travel of the balls. It is not that billiard players do the required mathematics, but rather that, if they were not playing in this mathematically optimal way, they would have been beaten too often and not become top billiard players in the first place.

The literature on service strategy originated with Gale (1971) and George (1973). Many articles followed. The typical approach is to assume two types of services, strong and weak, and then use summary statistics to verify whether the players have chosen the correct type. Players usually choose 'strong-weak', that is, the strong type as first service and the weak type as second service. But several times an alternative, such as strong-strong, seems better. We know from the discussion on regularity conditions in Table 9.1 and the substantial noise in summary statistics (rejection of hypothesis 9) that one should not take the results based on summary statistics too literally. The approach in this chapter, based on Klaassen and Magnus (2009), takes the noise issue seriously and allows for a continuum of services a player can choose (instead of just two).

Tennis players, when serving, have to make a number of strategic decisions, for example on the speed, the spin, and the direction of the service. Walker and Wooders (2001) examine whether tennis players aim their first service to the receiver's left or right in an optimal way. Their results corroborate optimality.

Pollard *et al.* (2010), Abramitzky *et al.* (2012), and Clarke and Norman (2012) investigate yet another strategy — regarding challenges. When should a player call upon the Hawk-Eye technology to arbitrate when he or she disagrees with the umpire's decision? About 3% of the points played are challenged, with a success rate of about 35%. Abramitzky *et al.* (2012) find that, in line with theory, players are more likely to challenge when the stakes are higher and when the (option) value associated with retaining one additional challenge is lower. For example, there should be, and there are, more challenges towards the end of a close set. On average, players tend to challenge too little, but their behavior is close to optimal when a simple model is assumed. This corroborates our conclusion on service strategy (page 149).

Why are tennis players so close to optimality? Because they can be and because they want to be. They can be close to optimality, because if they couldn't they would not play at Wimbledon in the first place. But there is more. Körding (2007) writes that the nervous system often comes up with near-optimal decisions in an uncertain world, and he uses tennis as an illustration. So it seems that the human nervous system is well-suited for optimization problems, of which serving and challenging in tennis are only simple examples.

Tennis players are also close to optimality because they want to be: winning a match at Wimbledon is very rewarding, not only in terms of money, but also in terms of world-ranking points and prestige. Lallemand *et al.* (2008) focus on money and demonstrate that larger prize differences between winning and losing a tennis match encourages players to increase effort. Another incentive comes from close competition. In uneven matches the gain from playing efficiently is low, as Figure 9.4 shows. But in close matches, players are expected to give the best they can. This is confirmed in Sunde (2009), who shows that tennis players exert less effort in a match where the difference in world rankings is large. In such a match the underdog often reduces effort because he or she will lose anyway, and the favorite reduces effort as well, because he or she will win anyway. This explains our finding on page 153 that a player serves more efficiently when the match is closer.

It is tempting — and in fact possible as the above studies on incentives and effort show — to relate results on tennis to other fields. In economics, we could consider payment schemes in firms. Do bonuses based on relative performance — higher bonus if you outperform your colleague — help to foster the average worker's effort, as economic theory predicts? The tennis outcomes by Lallemand *et al.* (2008) discussed earlier suggest that they do. Are bonuses less effective when the workers differ in quality? Applying Sunde's (2009) and our own tennis results, suggests that they are. Analysis on tennis can thus be useful to study economic problems that are difficult to address directly with traditional data. As a second example, the fact that tighter matches cause players to be more efficient suggests, more generally, that in a more competitive market firms are forced to be more efficient; otherwise they will be driven out of the market. This supports the view of many policy makers that measures aimed at strengthening market mechanisms lead to a more efficient economy. More examples are provided in Klaassen and Magnus (2009).

Examining tennis data helps not only to better understand economics, but also human behavior in general. Laboratory experiments show that humans make suboptimal choices. Such violations of optimality are often belittled by claiming that the incentives were insufficient or that the violations will be eliminated by learning or by market competition. Although Tversky and Kahneman (1986) agree that these factors are relevant, they question whether accounting for them would ensure fully optimal choices. In the end, this is an empirical issue. Our tennis data reveal that when incentives are high (Wimbledon), competition is fierce, and humans have learned a lot, then decisions indeed get closer to being optimal. In fact, the small deviations from full efficiency for top tennis players come close to an affirmative answer to Tversky and Kahneman's (1986) question of whether incentives, experience, and competition ensure fully rational choices.

Tennis is not the only sport that can help in understanding and predicting human behavior, and studies on baseball, basketball, and football (both American and soccer) have analyzed favoritism, market efficiency, optimal labor contracts, racial discrimination, and so on; see the references in Klaassen and Magnus (2009). Sports data are useful in these studies, because the players' objectives are clear, incentives to achieve them are strong, players are highly trained, and there is an abundance of high-quality data. Such circumstances are rare in psychology, economics, and related disciplines, thus making sports research relevant not only directly for the sport itself but also indirectly for the behavioral sciences. These advantages apply to tennis in particular: a researcher has to model only one player against one opponent (no team interactions), a good indication of quality exists due to the world rankings, the existence of two services doubles the information of a player's service strategy, and many points are observed in an objective way. Tennis is ideally suited for behavioral studies.

10 Within a match

So far we have considered matches between two players $\mathcal{I}$ and $\mathcal{J}$, and within each match we have taken all points served by $\mathcal{I}$ together and also all points served by $\mathcal{J}$. Many hypotheses, however, depend on the development *within* a match, because they concern special points or special games or 'momentum'. To test such hypotheses, the probability of winning a point on service must be allowed to vary over points. We therefore leave the analysis at match level behind and develop a model at point level. This point-level model will serve as our basis in this and the next two chapters.

In the current chapter we develop the point-level model and focus on two questions: whether points are independent and identically distributed (iid), as used extensively in Chapters 2–4; and whether new balls are an advantage to the server.

The idea behind the point model

The development of the point model closely follows the development of the match model, introduced in Chapters 6 and 7 and summarized on pages 115 and 116. In a match between $\mathcal{I}$ and $\mathcal{J}$, with $\mathcal{I}$ serving, the key element in the match model was that the probability p_i of winning a point was split up in an observed part β_i, capturing the impact of the players' rankings, and unobserved determinants collected in π_i. In formula,

$$p_i = \beta_i + \pi_i.$$

In the point model we wish to allow for variation in the winning probability over points, driven by some carefully chosen dynamic variables. At service point t of player $\mathcal{I}$, let d_{it} denote one such dynamic variable, reflecting for example what happened at the previous service point. (The word 'dynamic' signals that d_{it} is not constant over t.) If the impact of this dynamic variable is denoted by a parameter δ_0, then the match model can be extended to a point model by writing

$$p_{it} = \beta_i + \delta_0 d_{it} + \pi_i.$$

Writing p_{it} instead of p_i emphasizes that the probability may now vary over points.

We proceed in three steps. First, we ignore developments within a match ($\delta_0 = 0$) and only generalize the analysis from match to point level, thus obtaining the basic ingredients required for the model transition. Then we introduce dynamics by allowing δ_0 to differ from zero. And finally we allow for heterogeneity in the impact of dynamics, in order to be able to address such questions as whether momentum matters more or less for top players. In this final step we will substitute δ_0 by δ_i in the same way as we substituted β_0 by β_i. This will be our baseline model, and it will be used throughout this and the next two chapters.

From matches to points

To move from match-level to point-level analysis we first need to reconsider and generalize the moment conditions that determine the parameter estimates. We focus on the probability p_{it} of winning the tth service point. Our Wimbledon data set tells us whether player $\mathcal{I}$ wins or loses the tth service point, and we introduce the variable f_{it}, which takes the value one if the point is won and zero if the point is lost.

The first three moment conditions are similar to the ones in the match model:

$$\begin{aligned} \mathrm{E}(f_{it} - \beta_i) &= 0, \\ \mathrm{E}\left((f_{it} - \beta_i)(r_i - r_j)\right) &= 0, \\ \mathrm{E}\left((f_{it} - \beta_i)(r_i + r_j)\right) &= 0, \end{aligned}$$

where β_i, as before, depends on the ranking difference and the ranking sum via

$$\beta_i = \beta_0 + \beta_-(r_i - r_j) + \beta_+(r_i + r_j).$$

The conditions tell us that winning or losing a point on service depends on the rankings of the two players and on an unexplained part with mean zero, uncorrelated with the rankings.

The unexplained part consists of unobserved heterogeneity π_i between players and pure noise ϕ_{it}, which now also varies across points: at some points $\mathcal{I}$ is lucky, at other points $\mathcal{J}$. To filter out the noise we will use the variance of f_{it} and proceed in the same way as in Chapter 6 (page 97). The data are linked to the probability of interest p_{it} by

$$f_{it} = p_{it} + \phi_{it}.$$

Since we ignore dynamics in this section, the probability equation is simply

$$p_{it} = \beta_i + \pi_i.$$

Combining the two equations gives

$$f_{it} = \beta_i + \pi_i + \phi_{it}.$$

Because the unexplained part of f_{it} has zero expectation, the same holds for $\pi_i + \phi_{it}$, and we normalize again:

$$\mathrm{E}(\pi_i) = 0, \qquad \mathrm{E}(\phi_{it}) = 0.$$

The random noise ϕ_{it} is not correlated with the winning probability, so that

$$\mathrm{var}(f_{it}) = \mathrm{var}(\pi_i) + \mathrm{var}(\phi_{it}),$$

where $\mathrm{var}(\pi_i) = \sigma^2$ is the structural unobserved variation across players, that is, the unobserved variation excluding the variation in f_{it} arising from noise.

The noise variance $\mathrm{var}(\phi_{it})$ is restricted in the same way as before. In Chapter 6 there were T_i points underlying the frequency f_i. Here we look at just one point in f_{it}, so the restriction becomes that $\mathrm{var}(\phi_{it})$ equals the mean of $p_{it}(1-p_{it})$. In Chapter 6 we rewrote this restriction in terms of frequencies and divided by $T_i - 1$. We cannot do this here, because the number of points underlying the frequency is only one. Instead we write

$$\begin{aligned}\operatorname{var}(\phi_{it}) &= \mathrm{E}\left(p_{it}(1-p_{it})\right) = \mathrm{E}\left((\beta_i+\pi_i)(1-\beta_i-\pi_i)\right)\\ &= \beta_i(1-\beta_i)-\sigma^2.\end{aligned}$$

The next moment condition then becomes

$$\operatorname{var}(f_{it}) = \beta_i(1-\beta_i).$$

The final moment in Chapter 6 concerns the correlation. There, we only had to deal with the correlation between the two players $\mathcal{I}$ and $\mathcal{J}$ in the match. In the current model we also have correlation between all points served by $\mathcal{I}$ and all points served by $\mathcal{J}$. As before, we use ρ to denote the correlation between π_i and π_j. We assume that the random noise ϕ_{it} is uncorrelated with π_i and π_j, and also uncorrelated with noise at other points — luck is purely random. Under these assumptions we obtain

$$\begin{aligned}\operatorname{cov}\left(\pi_i+\phi_{it},\pi_i+\phi_{is}\right) &= \operatorname{var}(\pi_i) = \sigma^2 && (s\neq t),\\ \operatorname{cov}\left(\pi_i+\phi_{it},\pi_j+\phi_{js}\right) &= \operatorname{cov}(\pi_i,\pi_j) = \rho\sigma^2 && (i\neq j).\end{aligned}$$

In summary, the six moments for estimating the point model are:

$$\begin{aligned}\mathrm{E}(f_{it}-\beta_i) &= 0,\\ \mathrm{E}\left((f_{it}-\beta_i)(r_i-r_j)\right) &= 0,\\ \mathrm{E}\left((f_{it}-\beta_i)(r_i+r_j)\right) &= 0,\\ \operatorname{var}(f_{it}) &= \beta_i(1-\beta_i),\\ \operatorname{cov}\left(f_{it},f_{is}\right) &= \sigma^2 && (s\neq t),\\ \operatorname{cov}\left(f_{it},f_{js}\right) &= \rho\sigma^2 && (i\neq j).\end{aligned}$$

First results at point level

We have now moved from matches to points, but we have not yet introduced dynamics. Before doing so, let us compare the outcomes of the point model (without dynamics) with the outcomes of the match model used in previous chapters. We shall leave out tiebreaks from the point model — although they were included in the previous chapters — because the service rotation in a tiebreak is different than in an ordinary game and this complicates the analysis of the dynamics later in this chapter. As a result, all point models use 57,319 points for the men and 28,979 for the women.

	β_0 mean	β_- $r_i - r_j$	β_+ $r_i + r_j$	σ heterog.	ρ correl.
Men					
Point level data	64.6	9.0*	2.9*	5.6	−0.53
Match level data	64.9	8.0*	3.1*	5.0*	−0.44*
Women					
Point level data	55.9	17.1*	1.8*	4.9	−0.80
Match level data	56.3	15.9*	1.8°	4.2*	−0.79*

Table 10.1: *Explaining the probability of winning a point on service: point level versus match level*

The model comparison is provided in Table 10.1, which repeats the results from Table 7.2 at match level and confronts these with the new estimates at point level. The results at point level are different but not much different from the results at match level. All conclusions derived from Table 7.2 still hold. For example, we see that $\beta_- > \beta_+$ is still true at point level, so that the relative quality of the two players (quality difference) is more important than the total quality (quality sum). Also, the difference in strength between top players and lower-ranked players is greater in the women's singles than in the men's (larger β_- for thc women). If we are only interested in static (non-dynamic) results, then match data suffice. But if we wish to analyze dynamic questions, then we require point data and an appropriate statistical framework for analyzing them. Having obtained this framework we now move to the second step.

Simple dynamics

In the second step we consider the dynamic setup, be it in its simplest form, and introduce a dynamic variable d_{it} to capture variation in p_{it} as the match evolves. This variable is not deterministic, so that the moment conditions will become more complicated. The probability equation of winning service point t now reads

$$p_{it} = \beta_i + \delta_0 d_{it} + \pi_i,$$

where δ_0 represents the impact (assumed to be constant over all players) of the dynamic variable.

Testing iid

In Chapter 2 we introduced hypothesis 1: winning a point on service is an iid process. What this means is that the probability of winning a point is not affected by what happened at earlier points (independence) or by the characteristics of the point (the probability distribution is identical at each point). We have freely used the iid assumption throughout the book so far, claiming that it is a reasonable approximation. Now we have the tools to test it.

Let us split the iid hypothesis into two parts: one questioning whether points are independent, the other whether points are identically distributed. For the identically distributed part we take as our dynamic variable the importance imp_{it} of the current point:

$$d_{1,it} = imp_{it}.$$

Recall from the definition of importance (page 58) that it measures the impact of winning instead of losing point t on the probability of winning the match. To compute the importance of a point we need the point probabilities p_i and p_j. We don't know these probabilities, so they need to be estimated. Obvious candidates for the two estimators are $\hat{\beta}_i$ and $\hat{\beta}_j$. Because importance also depends on the score, it varies across points.

For the independence part we let the dynamic variable be the success or failure at the previous service point:

$$d_{2,it} = f_{i,t-1} = \begin{cases} 1 & \text{if previous service point was won,} \\ 0 & \text{if previous service point was lost.} \end{cases}$$

At the first point in a game the previous service point is not the point preceding the current point, and we set $d_{2,it} = 0$ at these points. (Alternatively we could have added a correction for the first point in a game, but it turns out that such a correction does not alter our results, so we prefer the simpler option.)

	Static determinants			Importance	Previous point
	β_0	β_-	β_+	δ_{10}	δ_{20}
	mean	$r_i - r_j$	$r_i + r_j$	constant	constant
Men	64.4	8.9*	2.8*	−9.8°	0.9*
	(0.4)	(1.0)	(0.7)	(6.6)	(0.4)
Women	56.0	17.0*	1.7*	−16.4*	1.2*
	(0.5)	(1.3)	(0.8)	(7.8)	(0.6)

Table 10.2: *Impact of importance and previous point on the probability of winning a service point, simple dynamics*

Using these two dynamic variables, our probability equation can be written as

$$p_{it} = \beta_i + \delta_{10} imp_{it} + \delta_{20} f_{i,t-1} + \pi_i,$$

and Table 10.2 contains the resulting estimates (standard errors in parentheses). We see that the static parameter estimates are hardly affected by including the dynamic variables: the estimates of β_0, β_-, and β_+ are almost the same as in Table 10.1. In future tables we shall therefore no longer report the β-estimates; neither do we report the estimates of σ and ρ.

If points are iid, then both δ_{10} and δ_{20} should be zero. Table 10.2 shows, however, that the importance estimate $\hat{\delta}_{10}$ is negative, which means that if a point is more important, then the server has less and the receiver more chance to win the point. The estimate is significant for women but not for men. The previous-point estimate $\hat{\delta}_{20}$ is positive, which means that if the previous point was won then there is a higher chance of winning the current point as well. Equivalently, if the previous point was lost, then the probability of winning the current point is lower. This suggests that swings are possible in p_{it}, which is an indication of the existence of 'momentum'. We shall have more to say on momentum in Chapter 12.

The estimate is significant for both men and women. Overall, the iid hypothesis 1 is rejected both in the men's and in the women's singles, so ups and downs do exist.

Why do players deviate from iid?

First, why do players deviate from the assumption that points are identically distributed? At important points a player may put more effort into the point, but if this is the case then it applies to both server and receiver, since the point is equally important to both of them. It is not clear why the receiver's extra effort would dominate the server's extra effort. A reason, perhaps, for the lower service-winning probability at important points is that players change their style of play at important points compared to ordinary points. If players deviate from their normal play and if the normal play is optimal, then they play suboptimally at important points. If, in addition, this suboptimality affects the performance of the server more than the performance of the receiver, then the receiver has an advantage at important points.

These are many 'ifs'. Let us test some of them.

Hypothesis 15: *Players play safer at important points.*

To test this hypothesis, we re-estimate our model but instead of taking 'winning the point' as the variable to be explained, we now consider 'first service in'. The estimates indicate that the impact of importance imp_{it} is positive, so more first services go in at important points. We also find fewer aces and fewer double faults at important points. We conclude that players serve more conservatively under pressure, in line with the hypothesis.

Next, why do they deviate from the assumption that points are independent? One possibility is that winning the previous point boosts self-confidence, leading to a winning mood and better play. This could lead to better performance using the same strategy, or it could lead to a different strategy. In the latter case, a player could take more risks than before, which would be bad under normal circumstances but could be good in different circumstances.

Hypothesis 16: *Players take more risks when they are in a winning mood.*

We test this hypothesis by considering the estimates of the impact of the previous point $f_{i,t-1}$. Winning the previous point reduces the probability of hitting the first service in, and increases the probabilities of an ace and a double fault. All this suggests that players do take more risks, thus supporting hypothesis 16.

Both identical distribution and independence are therefore rejected, because players change their style of play: they take less risks at important points, but more risks after winning a point. In addition, there may be a change in the probability of winning a point even when we abstract from the changes in the style of play.

Impact of deviations of iid

The rejection from iid is a potential problem for us (the authors), because several sections of this book, in particular Chapters 2–4, are based on the iid assumption. One possible defense of relying on the iid assumption is that the rejections from iid are not strong: the estimates are only about two standard errors away from zero, so the hypothesis is only marginally rejected. Since we have many observations and only a marginal rejection of the hypothesis, the distinction between significance and relevance, first discussed in Chapter 7, applies here.

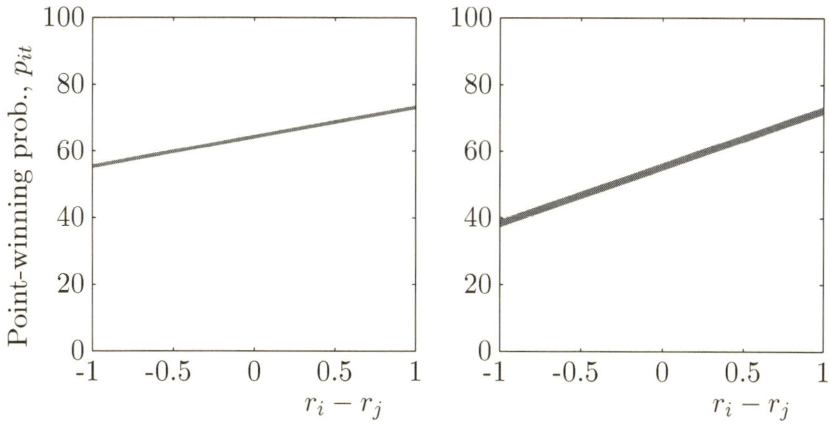

Figure 10.1: *Impact of importance on the probability of winning a service point for various ranking differentials (men left, women right)*

On the one hand we find that the estimated impact of importance is significant (at least for the women), from which we conclude that points are not identically distributed. On the other hand, if we plot the impact of importance in Figure 10.1, then we see that this impact is small. The figure plots the predicted p_{it} over the full range of $r_i - r_j$ for the case where the ranking-sum variable $r_i + r_j$ is set at its average, which is zero due to the centering. The predicted p_{it} is then given by

$$\hat{p}_{it} = \hat{\beta}_0 + \hat{\beta}_-(r_i - r_j) + \hat{\delta}_{10} imp_{it} + \hat{\delta}_{20} f_{i,t-1},$$

which we compute for a range of values of imp_{it} covering 95% of its values in the data set, while keeping $f_{i,t-1}$ fixed at its average. This produces a bundle of lines and we see that this bundle is narrow. Apparently, the predicted value of p_{it} is not much influenced by imp_{it}, even though the estimate of the associated parameter $\hat{\delta}_{10}$ is significant. The bundle is wider for the women than for the men, not because the parameter estimate is significant for the women, but because $\hat{\delta}_{10}$ is larger (in absolute value): -16.4 versus -9.8.

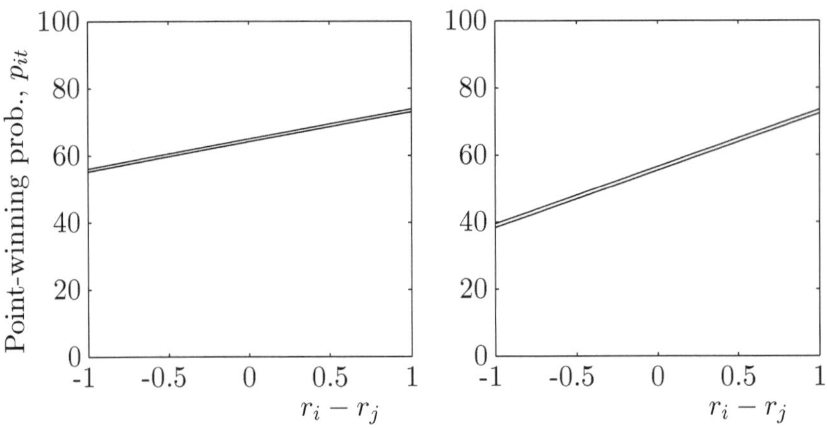

Figure 10.2: *Impact of previous point on the probability of winning a service point for various ranking differentials (men left, women right)*

Precisely the same story applies to the impact of the previous point, as Figure 10.2 illustrates. The same line is drawn as before, but now for $f_{i,t-1} = 1$ (previous point is won) and $f_{i,t-1} = 0$ (previous point is lost), while keeping imp_{it} fixed at its average. The resulting two lines are hardly distinguishable indicating that the impact of $f_{i,t-1}$ is small.

These two graphs and the analysis accompanying them illustrate again the important difference between significance and relevance. A deviation may be significant, while the impact of the deviation is negligible and irrelevant.

The small magnitude of the deviations from iid may be surprising to tennis fans. We can all think of matches where momentum seems to have swung from one player to the other. This applies, for example, to the Djokovic-Nadal Australian Open 2012 final, examined on page 96. At 1-1 in sets, the third set was easily won 6-2 by Novak Djokovic. It seemed that Rafael Nadal had lost his momentum. But did he? The answer is that we cannot exclude the possibility that the 6-2 third set score was just driven by chance. To show that deviations from iid *really* exist, chance has to be accounted for and after doing this, little evidence remains. Compare the discussion of the drop in Nadal's first service summary statistic from 71% (sets one and two) to 53% (third set), which we analyzed in Chapter 6. As there, we conclude that a seemingly big drop in performance may have been driven by chance only.

We conclude that the iid hypothesis 1 is rejected. Still, acting as if points are iid provides a reasonable approximation in many applications. This is important, because imposing iid simplifies many analyses, in particular those in Chapters 2–4.

The baseline model

There is no good reason why the impact of importance and momentum should be the same for all players. Therefore, in the third and final step of developing the dynamic model, we allow the parameters δ_{10} and δ_{20} to vary over players, and generalize them to

$$\delta_{1i} = \delta_{10} + \delta_{1-}(r_i - r_j) + \delta_{1+}(r_i + r_j),$$

and similarly for δ_{2i}, so that the impacts δ_{1i} and δ_{2i} depend on the rankings r_i and r_j, exactly mimicking the specification of β_i. Recall that $(r_i - r_j)$ and $(r_i + r_j)$ are centered, so that δ_{10} and δ_{20} represent the impacts for the average match. If $\delta_{1-} > 0$, then the negative impact of imp_{it} via δ_{10} is somewhat reduced for servers who have a better ranking than their opponents. If $\delta_{2+} < 0$, then the positive impact of $f_{i,t-1}$ via δ_{20} is somewhat reduced for servers in a top match.

This gives our baseline model, with probability equation

$$p_{it} = \beta_i + \delta_{1i} imp_{it} + \delta_{2i} f_{i,t-1} + \pi_i.$$

If we let $\delta_0 = \delta_- = \delta_+ = 0$ for both variables, then we obtain the setup in the first step (no dynamics), and if we let $\delta_- = \delta_+ = 0$, then we obtain the setup in the second step (simple dynamics).

The baseline model will be our main tool in the remainder of this book. In most applications of the model we will add *one* further dynamic variable to the model, corresponding to one specific hypothesis under consideration. Only at the end of Chapter 12 shall we add more than one variable to the baseline model.

	Importance			Previous point		
	δ_{10} mean	δ_{1-} $r_i - r_j$	δ_{1+} $r_i + r_j$	δ_{20} mean	δ_{2-} $r_i - r_j$	δ_{2+} $r_i + r_j$
Men	−13.0°	31.5°	42.6*	0.9*	0.0°	−2.2*
Women	−16.7*	116.3*	11.7°	1.2*	3.3°	−4.2*

Table 10.3: *Baseline model: impact of importance and previous point on the probability of winning a service point*

The estimates of the baseline model are given in Table 10.3. These confirm the rejection of iid in hypothesis 1, so that the winning probability p_{it} does indeed vary over points. Formally, this is a test on the joint hypothesis that all six δ-parameters (three for each variable) are zero. The lack of independence and the lack of identical distribution both matter in this rejection.

The main contribution of the table concerns the impact across players. Player homogeneity can be investigated by testing the joint hypothesis that rankings are irrelevant, both for the impact of importance and for the impact of the previous point, so that the model of the previous section suffices. The joint test rejects that δ_{1-}, δ_{1+}, δ_{2-}, and δ_{2+} are all zero. Deviations from iid are therefore player-dependent.

When we compare the δ-estimates of men and women, we see that the signs are the same, which shows that the impacts of importance and previous point work in the same direction in the men's singles and in the women's singles. Equality of the δ-parameters for these two independent samples is not rejected.

Top players and mental stability

Players are trained by their coaches to be mentally stable, forget about the score, forget about the past, and focus on the current point. In other words, they are trained to play points as iid-like as possible. But maybe not all professionals are equally successful in putting this advice into practice.

Hypothesis 17: *Top players are more stable than others.*

We study player heterogeneity for both importance and previous point in more detail, first importance and then previous point.

Importance

Regarding importance, all estimates in Table 10.3 point in the same direction, although not all of them are significant. When stakes are high and points important, the server has a disadvantage: $\delta_{10} < 0$. This disadvantage is weaker when the server is a better player than the receiver ($\delta_{1-} > 0$). In more detail, if the quality of player $\mathcal{I}$ (measured by r_i) increases, then he or she will perform better at important points (δ_{1i} less negative). Similarly, if the quality of opponent $\mathcal{J}$ increases, then the opponent will perform better at important points, so that $\mathcal{I}$ will perform worse (δ_{1i} more negative). Hence the players' rankings work against each other, so that their impact on δ_{1i} depends on the difference of the rankings $(r_i - r_j)$.

If δ_{1+} were zero, then the above explanation would imply that the quality of server $\mathcal{I}$ matters as much (in absolute value) as the quality of receiver $\mathcal{J}$. However, the service dominance in tennis suggests that the quality of the server should matter more. This explains why we find $\delta_{1+} > 0$, because this makes the total impact of r_i (measured by $\hat{\delta}_{1-} + \hat{\delta}_{1+} = 74.1$ in the men's singles) stronger in absolute value than the impact of r_j (measured by $-\hat{\delta}_{1-} + \hat{\delta}_{1+} = 11.1$ in the men's singles). For the women, the sum of the rankings has a smaller impact, which is consistent with the fact that service dominance is larger for men than for women.

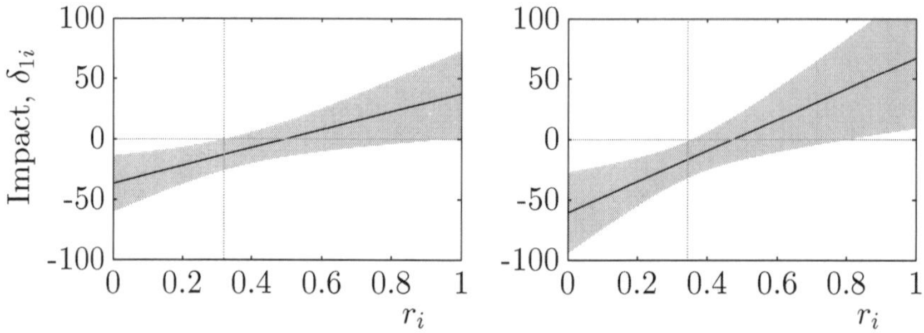

Figure 10.3: *Relevance of the server's ranking for the impact of importance (men left, women right)*

To better understand the relevance of the server's ranking for the impact of importance, let us consider a server $\mathcal{I}$ playing against an average opponent $\mathcal{J}$. The average opponent will have $r_j = 0.32$ for the men and $r_j = 0.34$ for the women, corresponding to world rankings 43 and 39, respectively. Figure 10.3 illustrates how the estimated impact δ_{1i} of importance depends on the ranking r_i. We see the estimate (solid line) and its uncertainty (confidence interval), indicating that with 95% certainty the true δ_{1i} lies within the band. The vertical reference line is placed at $r_i = r_j$, the average. Because the opponent is an average player, δ_{1i} at the point $r_i = r_j$ is equal to the estimated δ_{10}, that is, -13.0 (just insignificant) for the men and -16.7 (just significant) for the women. The borderline significance corresponds to the conclusion, earlier in this chapter, that the deviation from iid is not strong. Still, the fact that two independent samples — men and women — yield such similar results is reassuring: the importance of a point matters.

The difference in δ_{1i} between the top player with $r_i = 1$ and a weak player with $r_i = 0$ is $\delta_{1-} + \delta_{1+}$, which is 74.1 for the men and 128.0 for the women, both significant. So if a top player serves at an important point against the average opponent, he or she wins more often than a weaker player, not only because the stronger player wins more points anyway, but also *because he or she is more successful at important points.*

Is this because the top player can raise his or her level, or because the weaker player underperforms, or both? In Figure 10.3 the opponent $\mathcal{J}$ is always the average player, while the quality of the server, player $\mathcal{I}$, varies. We consider three servers: average, top, and weak, respectively. If server $\mathcal{I}$ is an average player, then we observe a lower winning probability p_{it} at important points. If player $\mathcal{I}$ underperforms as a server, then he or she will also most likely underperform as a receiver. Because players $\mathcal{I}$ and $\mathcal{J}$ are the same (more precisely, have the same quality), this implies that $\mathcal{J}$ also underperforms. It could also be that both $\mathcal{I}$ outperforms and that $\mathcal{J}$ outperforms even more, but because the server's performance most likely dominates the receiver's in determining p_{it}, this is not a credible scenario. Hence, both $\mathcal{I}$ and $\mathcal{J}$ underperform.

Next, consider the top server (at the right side of the graph) at important points. We know that the opponent, the average player, underperforms, which by itself increases p_{it}. So if the top player outperforms, p_{it} would increase even more. This does not correspond to Figure 10.3, so the top player does not outperform. Underperformance by the top player is not credible, because this underperformance would then dominate the underperformance of the receiver, leading to a lower p_{it}, contradicting the figure. The top player thus plays at his or her normal level at important points.

For the weak server (left side of the graph) we find a reduction in p_{it}. Because the opponent underperforms, which by itself raises p_{it}, the weak server apparently substantially underperforms. We conclude that below-average players underperform at important points, and that above-average players are not much affected by importance. *Top players do not perform better, but lower-ranked players perform worse, at important points.*

Now we can address the identical distribution component of hypothesis 17, which states that top players are more stable than weaker players. From the previous discussion we conclude that for top players we do not reject identical distribution, but for all men and women below ranking 50 ($r_i = 0.3$) we do reject identical distribution, thus supporting hypothesis 17.

The previous point

Player heterogeneity also matters for the influence of the previous point on the current point. Table 10.3 confirms our earlier conclusion that winning the previous point increases the probability of winning the current point ($\delta_{20} > 0$) and that losing the previous point decreases this probability, both for men and for women. The new result is that the stronger the players the weaker the effect ($\delta_{2+} < 0$). There is no evidence that the ranking difference matters ($\delta_{2-} = 0$). It makes sense that only the sum of the rankings matters, because if a stronger player makes the match more stable (less dependence), then this holds for both the server and the receiver. High values of r_i and r_j now reinforce each other, and this is captured by the ranking sum but not by the ranking difference. If two top players meet then the match will be stable, but when two weaker players meet then there will be more performance swings. (When two amateurs play, there will be many swings, although this is only based on casual observation without any scientific foundation.)

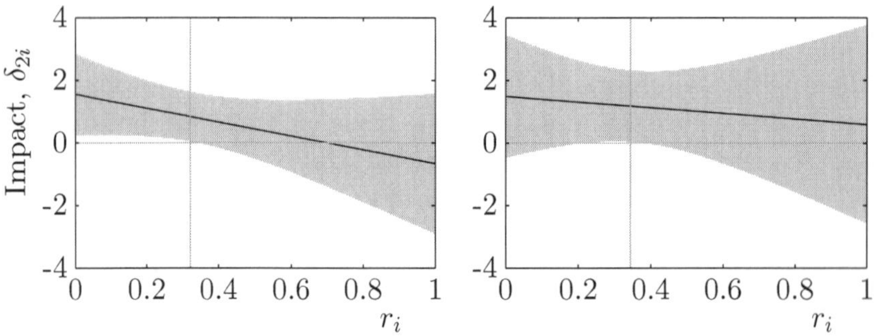

Figure 10.4: *Relevance of the server's ranking for the impact of the previous point (men left, women right)*

How much dependence is left for the top players? Do they succeed in switching off the impact from the previous point completely and 'forget about the past'? In Figure 10.4 we plot the estimated δ_{2i} against r_i for the average r_j. For the average server the estimated δ_{2i} is just significant for both men and women, so that winning (losing) the previous point has a positive (negative) impact on the current point. Below-average players also exhibit such ups and downs, but above-average players (including the very top players) play the points independently.

Lessons from the baseline model

Our conclusion is that hypothesis 17 is supported. Top players not only follow a more efficient service strategy in the match as a whole (see page 153), but they are also more stable across points within a match.

Points in men's and women's tennis are not iid, but the deviation from iid is not large, and iid will be a good approximation in many directions. Players perform worse on service at important points, but this performance dip is smaller for better players. Performance at the previous point carries over to the current point, but above-average players are able to avoid this effect. For top players we do not reject iid; they are stable. But below-average players deviate from iid in two ways: their performance in the current point depends on what happened in the previous point and they are also affected by the importance of a point.

New balls

Tennis as a game has a long history which goes back to the Greeks and Romans. But it was not until 1870 that it became technically possible to produce rubber balls which bounce well on grass. When The All England Lawn Tennis and Croquet Club decided to hold their first championships in 1877, a three-man subcommittee drew up a set of laws. Rule II stated that

> the balls shall be hollow, made of India-rubber, and covered with white cloth. They shall not be less than 2 1/4 inches, nor more than 2 5/8 inches in diameter; and not less than 1 1/4 ounces, nor more than 1 1/2 ounces in weight.

The quality of the tennis balls has gradually improved, and currently various types of balls with well-defined characteristics exist. From 1881 to 1901 the balls were supplied by Ayres; thereafter by Slazenger and Sons. Yellow balls were introduced at the hundredth anniversary of the Wimbledon championships in 1986. During the 1877 championships 180 balls were used; now more than 50,000.

In 2002 the ITF changed the ball approval from one ball for all surfaces to three ball types: a faster ball (less compression) designed to speed up play on slow courts (like clay), a slower ball (slightly larger) designed to reduce the service advantage on fast surfaces (like grass), and the original ball for medium surfaces (like acrylic). This freedom to use different types of balls was not a success. Players objected to it and the experiment was discontinued. All tournaments currently use the medium ball, but there is some evidence that tournaments on slow courts prefer balls at the faster end of the medium, while tournaments with fast surfaces prefer balls at the slower end. Slower balls are also recommended for tennis played at high altitude (above 4000 feet, 1219 meters) as an alternative to the pressureless ball.

During a grand slam tennis match new balls are provided after the first seven games (to allow for the preliminary warm-up) and then after each subsequent nine games. Most commentators and many spectators believe that new balls are an advantage to the server. But are they right?

Hypothesis 18: *New balls are an advantage to the server.*

To find out, let us consider Figure 10.5. The age of the balls is indicated from 0 (new balls) to 8 (old balls). The age is 0 in the game with new balls, 1 in the next game, and so on. After the game where the age of the balls equals 8, new balls are used so that the age is again 0. During the five minutes of warming up before the match begins, the same balls are used as in the first seven games. Thus it makes sense to set the age of the balls in the first game of the match at 2.

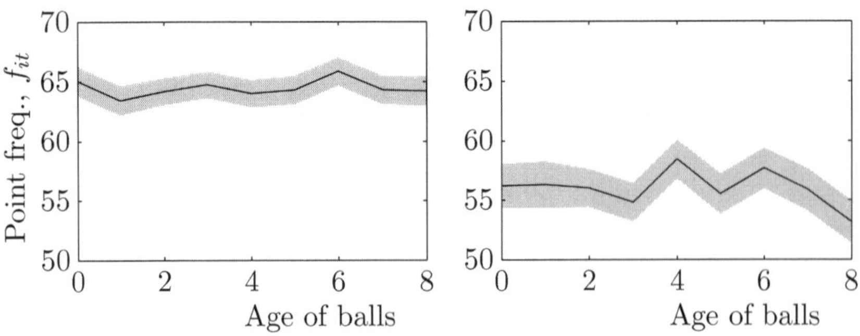

Figure 10.5: *Percentage of service points won depending on the age of the balls (men left, women right)*

If the hypothesis that new balls provide an advantage were true, the dominance of service, measured by the probability of winning a point on service, would decrease with the age of the balls. The figure does not support this hypothesis, at least not in the men's singles. For the women, the relative frequency of winning a point on service with balls of age 8 is 53.3%, hence lower than with balls of age 0 where the relative frequency is 56.2%. This drop in the relative frequency of winning a service point suggests that the quality of the balls drops as well. But if we add confidence intervals, then we see that the drop for the women may well be random noise. The drop is not significant, and hence there is no statistical support for the hypothesis.

The above analysis is graphical. To make the analysis statistically more sound, we call on our baseline model, which includes both importance and previous point, and add to this model one further dynamic variable. We consider two possible choices for this additional variable to test hypothesis 18. The first is

$$d_{3,it} = \begin{cases} 1 & \text{in games where new balls are used,} \\ 0 & \text{in all other games.} \end{cases}$$

This is simple: it just distinguishes new balls (age = 0) from no new balls (age > 0), and focuses on the shock that switching from old to new balls creates. The question is whether there is any effect on the winning probability.

	New balls			Age of balls		
	δ_{30} mean	δ_{3-} $r_i - r_j$	δ_{3+} $r_i + r_j$	δ_{40} mean	δ_{4-} $r_i - r_j$	δ_{4+} $r_i + r_j$
Men	0.7°	−4.2*	−1.4°	0.0°	0.4°	−0.2°
Women	−0.1°	3.7°	5.4*	−0.1°	−0.5°	−0.5°

Table 10.4: *Impact of new balls and age of balls*

The left half of Table 10.4 provides the estimates of the baseline model with d_3 as an added variable. The estimates corresponding to d_1 and d_2 have been omitted because they are quite stable and do not change much compared to Table 10.3. In Table 10.4

most estimates are insignificant (recall that the symbol $^\circ$ denotes insignificance), and the two significant estimates (denoted by *) do not support the hypothesis. We conclude therefore that there is no support for hypothesis 18.

Next, we introduce an alternative dynamic variable to test the same hypothesis,

$$d_{4,it} = \text{ age of the balls } (0, \ldots, 8).$$

The variable d_4 is more sophisticated than d_3, because it distinguishes between the age of the balls in a gradual manner to capture the fact that the wear and tear of the balls occurs gradually. The right half of Table 10.4 presents the estimates of the baseline model with d_4 as an added variable. Nothing is significant, so hypothesis 18 is not supported by the data.

Is there then no difference between new balls and old balls? Yes, there is. New balls, just out of the can, are smooth and bouncy, whereas older balls are softer and fluffier. The older balls provide more grip, making it easier to control the service. More first services will go in and fewer double faults occur. On the other hand, the smooth new balls travel faster through the air, so that if the first service is in, it will have a higher chance of winning the point.

Thus we have two forces working against each other: the probability that the service is in (x in the notation of Chapter 9) decreases when balls are new, but if the service is in, the probability of winning the point (y) increases. What matters is the product xy, and whether this increases or decreases is an empirical issue. We find that the two forces balance each other, so that no new-ball advantage can be found from the data.

Further reading

An extensive literature exists on the question of whether points in sports (not only tennis) are identically distributed, and in particular whether they are independent. 'Streaks' in baseball and the 'hot hand' in basketball are examples of dependence. We will discuss the literature on identical distribution and independence separately in the next two chapters. Here we focus on iid in general and on the new-balls hypothesis in particular.

The analysis of iid in the current chapter follows Klaassen and Magnus (2001), which is a rather technical paper discussing the estimation method more fully than we do here. Newton and Aslam (2006) start from the assumption that there exist deviations from iid. They then model these deviations, simulate their impacts, and report that, even when relatively strong non-iid effects are introduced, results derived from iid are remarkably robust and accurate. This is good news for analyses based on the iid assumption, as in the early chapters of this book.

Our explanations for deviations from iid — safer play at important points and riskier play in a winning mood — are more advanced versions of what we reported in Magnus and Klaassen (1996). In this earlier paper we found evidence that players at breakpoint opt for safety (more first services in and fewer aces), that after an ace players increase risk (fewer first services in and more aces), and that after a double fault players take less risks (more first services in and fewer aces). These findings are all in line with our explanations for iid deviations.

A different type of analysis is provided by Paserman (2010), in the context of identical distribution. Paserman uses stroke-by-stroke data from grand slam tournaments played in 2006 and 2007, and finds that at important points, both for men and women, rallies last longer, because players hit fewer winners and make fewer unforced errors. He attributes this to a more conservative and less aggressive playing strategy, and shows that this affects servers more than receivers. Paserman's findings support our claim that the suboptimality caused by too conservative play is larger for the server than for the receiver.

Strategic adjustments at important points and in a winning mood derived from tennis data also have implications outside tennis, for example in economics. If salaries of agents working in the financial sector would contain a bonus *and also* a substantial malus component, then the consequences of their activities would matter in both directions (like winning *or* losing a tennis match). The behavior of professional tennis players then suggests that top financial agents will pursue safer actions, reducing the possibility of a banking crisis. Tennis provides a clean environment to test this theory.

The quality of the balls matters in professional tennis. Goodwill *et al.* (2004) simulate trajectories of new (0 impacts) and heavily worn (1500 impacts) balls, and find that after a typical first serve the worn ball lands about fifty centimeters farther than the new ball, because the worn ball does not 'dip' as much during the flight. Strict rules exist, available at the website of the ITF Technical Department, to ensure that the changes in the balls do not have too much influence on play. The ITF tests balls in their Technical Centre, a world-leading tennis testing laboratory; see Spurr and Capel-Davies (2007). Despite these strict technical rules, there could still be a new-ball effect. Our results in the current chapter show, however, that no effect on the probability of winning a point on service can be found, in line with Magnus and Klaassen (1999a) and Norton and Clarke (2002). These two papers provide more details, for example regarding the different dimensions of winning a service point, such as hitting the first service in and winning the point given that the first service is in.

11

Special points and games

We shall employ the baseline model developed in the previous chapter to test some often-heard hypotheses relating to whether or not points are identically distributed or independent or both. The next chapter addresses independence, the current chapter focuses on the issue of identical distribution. There may be more than just differences in importance, studied in the previous chapter, that make players perform differently across points. Perhaps breakpoints are particularly special or serving first in a set matters.

Big points

Big points are points that are especially important, such as breakpoints. It is often claimed that top players perform particularly well on those points. If true, top players would have a double advantage. Not only are they better players, but they can also raise their game when it really matters. We formulate this hypothesis as follows.

Hypothesis 19: *Real champions win the big points.*

Let us first think of real champions as seeds and of big points as breakpoints, a drastic simplification. When a seed serves in the men's singles, the probability of winning a point on service is 67.9% at breakpoint down and 68.7% at other points (seeded and non-seeded receivers taken together). The difference is not significant. However, when a non-seed serves, the difference is significant: 59.3% at breakpoint and 63.5% at other points. Therefore, real champions perform better at breakpoint down than other players.

However, real champions do not win more points at breakpoint down than usual. It is the non-seed who wins fewer points.

The situation is different for the women, where we find no evidence for our hypothesis. In the women's singles, the probability of winning a point on service is 58.2% at breakpoint down and 61.7% at other points for a seed (seeded and non-seeded receivers taken together). For a non-seed these numbers are 50.3% at breakpoint down and 54.8% otherwise. Both differences are significantly negative, and the decrease for the seeds (3.5%-points) is only marginally smaller than the decrease for the non-seeds (4.5%-points). So, real champions do not perform better than weaker players in the women's singles.

So far we have defined the word 'champion' by only looking at the server: if the server is a seed, then he or she is a champion. What if the receiver is also a seed? As in Chapter 8 we encounter the problem that only considering the server is too simplistic.

Therefore, as in Chapter 8, we shall distinguish between an absolute and a relative interpretation of the word 'champion'. In an absolute sense, a player is a real champion if he or she is very good (for example a seed) irrespective of the opponent in the match. In a relative sense, a player is a real champion if he or she is much better than the opponent, say a seed in a match against a non-seed. Because an absolute champion is typically a better player than the opponent, characteristics of relative champions also matter for absolute champions. To exclude this overlap, we again define a cleaned version of absolute champion (corrected for quality difference), and then consider matches between two absolute champions, for example two seeds.

Given these finer distinctions we again consider the question of whether champions play their best tennis at breakpoints, and Table 11.1 provides some of the answers. Regarding absolute champions corrected for quality difference, we compare matches between two seeded players with matches between two non-seeded players. In the men's singles, a seed playing against another seed will perform worse at breakpoint down (65.3% against 66.9%), and the same is true for two non-seeds (60.0% against 64.2%). The difference is bigger for the non-seeds (−4.2 versus −1.6), although not significantly so. This is also true in the women's singles (−5.4 versus 2.1), but there the difference is significant. Hence, there is mild

	Sd-Sd	Sd-NSd	NSd-Sd	NSd-NSd	Total
Men					
At breakpoint	65.3	68.9	57.7	60.0	60.9
At other points	66.9	69.3	61.3	64.2	64.8
Difference	-1.6°	-0.4°	-3.6^{*}	-4.2^{*}	-3.9^{*}
Women					
At breakpoint	58.7	57.9	48.6	51.0	51.9
At other points	56.6	63.5	50.2	56.4	56.6
Difference	2.1°	-5.6^{*}	-1.6°	-5.4^{*}	-4.7^{*}

Table 11.1: *Percentage of points won on service at breakpoints and other points, seeded (Sd) and non-seeded (NSd) players*

confirmation of the hypothesis.

In a relative sense, if the hypothesis were true, a seeded server would perform better against a non-seed (Sd-NSd) than a non-seeded server against a seed (NSd-Sd) at breakpoint compared to other points. For the men the seed does perform better, but the difference of 3.2%-points (−0.4 minus −3.6) is not significant. For the women the difference between −5.6 and −1.6 is also not significant. So, we find no evidence for our hypothesis in a relative sense.

The different interpretations of 'champion' lead to mixed results. Maybe this is caused by the fact that our approach is too simple. Perhaps the quality measure is too simple: we distinguished only between Sd and NSd. If a breakpoint occurs in a NSd-NSd match, then — given that there are strong and weak non-seeds — the breakpoint may simply indicate that the server is a weaker player than the receiver, so that it is quite natural that breakpoints are more often won by the receiver. This should not be interpreted as support for hypothesis 19. We encounter here another example of sample selection bias (see page 81): weaker servers are overrepresented in the 'at breakpoint' sample, and this reduces the estimated winning probability. This bias cannot be repaired by only considering non-seeded players; a proper quality measure is required.

Another simplification, and thus another possible reason for the mixed results, is that we distinguished only between breakpoints and non-breakpoints, where all breakpoints are considered equally important and all other points equally unimportant. Here too we need a finer measure, and this finer measure is available: importance.

Big points and the baseline model

Suppose we generalize the previous model by using our quality variable based on expected round (see page 110) instead of seeded versus non-seeded; and our importance measure instead of breakpoints versus non-breakpoints. This would be an improvement, but even then the occurrence of a breakpoint could indicate that the receiver is in a winning mood. This would also explain underperformance at breakpoint down. What is required is a model that properly accounts for quality, does not split the sample, accounts for momentum, and better captures the differences in importance across points. This is precisely what our baseline model attempts to achieve.

The baseline model corrects for quality and contains importance (d_1) and previous point (d_2). To these two variables we add a breakpoint dummy

$$d_{5,it} = \begin{cases} 1 & \text{if current point is a breakpoint,} \\ 0 & \text{otherwise.} \end{cases}$$

The extended model thus contains d_1, d_2, and d_5.

	δ_{50} mean	δ_{5-} $r_i - r_j$	δ_{5+} $r_i + r_j$
Men	-1.3°	1.9°	-2.3°
Women	-2.6^{*}	-5.7^{*}	0.1°

Table 11.2: *Impact of breakpoints*

We report the estimates of this model in Table 11.2, but only for d_5 because the estimates of the other parameters are roughly similar to the ones reported before. Recall from page 130 that the sum ($\delta_- + \delta_+$) captures absolute champions, while δ_+ captures the absolute version corrected for quality differences and δ_- captures relative champions.

What do we find? In the men's singles we know from Table 10.3 (page 172) that absolute quality matters for the impact of importance ($\delta_{1-} + \delta_{1+} > 0$). Given this influence, Table 11.2 reveals that the occurrence of a breakpoint does not have any further impact: the estimate of $\delta_{5-} + \delta_{5+} = -0.4$, not significant. The same holds for the women where the estimate is -5.6, also not significant. If we account for quality differences, then the values of δ_{5+} reveal again that breakpoints have no additional impact, either for men or for women. Regarding relative champions (δ_{5-}) the occurrence of a breakpoint does have further impact, but only for women.

What then is our conclusion regarding hypothesis 19? Is it true that real champions win the big points or not? Absolute champions perform better at the big points than other players, not because they play better tennis than usual but because their opponents play worse.

Serving first revisited

In Chapter 2 we briefly discussed hypothesis 2: it is an advantage to serve first in a set. Under the assumption that points are iid — the assumption underlying Chapter 2 — the hypothesis is rejected on theoretical grounds. Under iid, it makes no difference whether one serves first in a set or not. However, we have just seen that the iid assumption needs to be rejected for below-average players. This does not necessarily imply that it now makes a difference who serves first in a set, but it does mean that we can no longer reject the hypothesis on theoretical grounds. We should test it empirically.

We start again simply by looking at basic statistics from the data, not using our statistical model. Our first calculations seem to indicate that hypothesis 2 must be wrong. Overall only 48.2% of the sets played in the men's singles (50.1% in the women's singles) are won by the player who serves first in the set. Neither of the two percentages is significantly different from 50%.

Table 11.3 adds a little more detail by looking at individual sets. In the men's singles, the estimated probabilities of winning a set when serving first is 55.4% in the first set, and 44.3%, 43.5%,

	Player	Winner of previous set		
	serves first	serves first	receives first	difference
Men				
Set 1	55.4	–	–	–
Set 2	44.3	72.5	68.0	4.5°
Set 3	43.5	73.9	72.1	1.8°
Set 4	51.0	62.9	60.2	2.7°
Set 5	48.8	48.3	51.0	−2.7°
Women				
Set 1	56.6	–	–	–
Set 2	44.0	72.0	75.2	−3.2°
Set 3	47.8	63.5	60.1	3.4°

Table 11.3: *Percentage of sets won depending on whether a player serves first or receives first in a set*

51.0%, and 48.8% in the subsequent sets. In the women's singles, the estimated probabilities of winning a set when serving first is 56.6% in the first set, and 44.0% and 47.8% in the second and third sets. These numbers suggest that it may be an advantage to serve first in the first set, but that this advantage becomes a disadvantage in the second and subsequent sets.

Surely this is peculiar. One may doubt that there is an advantage in serving first, but if this is not true why should there be an advantage in receiving first, as there seems to be in the second and subsequent sets? The answer to this little puzzle is as follows. The player who serves first in a set, if it is not the first set, is usually the weaker player. This is so because of a combination of two factors. First, it is likely that the stronger player won the previous set; and second, it is likely that the last game of the previous set was won by the server. As a result, the loser of the previous set typically serves first in the next set. He or she is then more likely to lose the current set, not because of a (dis)advantage to serving first, but because he or she is the weaker player. The probabilities in the second and subsequent sets *must* be lower than 50%, and the estimated probabilities reflect this. We have failed to correct for quality.

To correct for quality and provide a more credible analysis, we employ conditional rather than unconditional probabilities. Thus we consider players who have won the previous set, and compare the estimated probability that they win the current set when serving first with the probability that they win the current set when receiving first. The estimates of these conditional probabilities are also provided in Table 11.3. The difference between the two probabilities (serving and receiving) is sometimes estimated as positive, sometimes as negative, but never significant. So now we conclude that there is no service advantage in the second and subsequent sets.

Perhaps there is, however, a possible advantage in the *first* set. We shall consider this possibility in the next section, but before doing so we confront hypothesis 2 with our baseline model. We therefore extend the baseline model with the variable

$$d_{6,it} = \begin{cases} 1 & \text{if player } \mathcal{I} \text{ served first in the current set,} \\ 0 & \text{if player } \mathcal{I} \text{ received first in the current set,} \end{cases}$$

and this leads to the following estimates, again deleting the estimates for d_1 and d_2.

	δ_{60} mean	δ_{6-} $r_i - r_j$	δ_{6+} $r_i + r_j$
Men	-0.5°	-0.2°	1.7°
Women	0.4°	1.3°	-6.4^{*}

Table 11.4: *Impact of serving first in a set*

Five of the six parameter estimates in Table 11.4 are insignificant. What matters most is that the *joint* hypothesis

$$\delta_{60} = \delta_{6-} = \delta_{6+} = 0$$

is not rejected, either for the men or for the women. In this case, the more sophisticated analysis (using the full strength of the baseline model) confirms the less sophisticated analysis (using only match frequencies). In the less sophisticated analysis we corrected for quality by the noisy 'winner of the previous set' indicator. In the baseline model the quality correction is much better, but the conclusion is the same: there is no evidence of an advantage of serving first in a set, except perhaps in the first set.

The toss

Is the first set different from the other sets in the sense that there is advantage in serving first? If so, what explains it? Suppose there is such an advantage in the first set. Then, if a player wins the toss, he or she should elect to serve.

Hypothesis 20: *The winner of the toss should elect to serve.*

We first examine this hypothesis using Table 11.5. The estimated probability of winning a service point, not in the first set, is 64.4% in the men's singles and 55.6% in the women's singles. The probability of winning a service point in the first set (excluding the first two games) is about the same, namely 64.3% (men) and 56.4% (women). The differences −0.1 and 0.8 are not significant, as indicated by the symbol °.

	Men		Women	
	%	diff.	%	diff.
1st game in match	67.8	3.4*	60.4	4.8*
2nd game in match	64.8	0.4°	53.4	−2.2°
1st set except games 1 and 2	64.3	−0.1°	56.4	0.8°
Match except 1st set	64.4	–	55.6	–

Table 11.5: *Percentage of winning a service point at various stages of the match (diff = difference with respect to match except 1st set)*

The probability that points in the very first game of a match are won by the server is higher, namely 67.8% (men, significant) and 60.4% (women, also significant). The reason why the set-winning probability in the first set is higher is entirely due to the first-game effect. Only the first game matters, not the first two games, because in the second game the effect has already disappeared as the second line in Table 11.5 shows.

Hypothesis 20 thus appears to be true. In general, if a player can choose, he or she should elect to serve first in a match, because the very first game of the match is seldom lost by the server and this causes a first-set advantage. In fact, this strategy is common among professional players, although choosing to receive is no exception, generally a mistake according to our simple analysis.

There is, however, another possibility. Maybe strong players elect to serve after winning the toss, while weaker players (not always, but more frequently than stronger players) elect to receive. If this is what happens in practice, than naturally the first game is won more often by the server than on average, not because of a first-game advantage, but because the stronger player serves. In that case, we would then also expect to see a negative effect in the second game, where the weaker player serves, but such an effect does not appear to be present. Hence, there is no indication that strong players, after winning the toss, elect to serve more often than weaker players.

Less simplistic and more informative is the approach relying on our baseline model. This time we add the dynamic variable

$$d_{7,it} = \begin{cases} 1 & \text{if service point occurs in first game of match,} \\ 0 & \text{otherwise,} \end{cases}$$

to our baseline variables d_1 and d_2. The impact of this additional variable is given in Table 11.6.

	δ_{70} mean	δ_{7-} $r_i - r_j$	δ_{7+} $r_i + r_j$
Men	3.3^*	0.2°	-3.2°
Women	3.8^*	-6.7°	-3.0°

Table 11.6: *Impact of serving in the first game of the match*

The estimate of the impact δ_{70} is significant and positive, both for men and for women. The first-game bonus does not depend on the rankings (the estimates of δ_{7-} and δ_{7+} are not significant).

The winner of the toss, irrespective of his or her quality, is apparently rewarded with a 3- to 4%-points higher probability of winning a point in the first game, provided he or she elects to serve. At game level, this advantage is 6- to 8%-points, using the formula on page 15 and taking 60% as the normal point-winning probability.

The fact that the first game of the match is different from other games raises the question why this should be the case. We can only guess. Maybe the receiver, rather than trying hard to win points in the first game, learns to read the server's strategy, judges his or her strengths and weaknesses, and settles down. But why the receiver in the second game is less passive remains a mystery.

Further reading

The existing literature on testing the identical distribution of points focuses on the impact of importance and breakpoint on the point-winning probability, corresponding to the real champions hypothesis 19. The literature is not in agreement, and the differences between the outcomes of the various approaches emphasize again the importance of a good quality correction.

O'Malley (2008) compares the performance of the server at breakpoint down to the performance at other points, without correcting for quality, and he reports that servers underperform at breakpoint. Magnus and Klaassen (1999b, 2008) do the same using the simplistic quality correction based on seeded and non-seeded players, and they find the mixed results of Table 11.1. A better quality correction is proposed in González-Díaz *et al.* (2012), although they still ignore unobserved quality. A second improvement is that they replace the breakpoint dummy by the importance variable *imp*. With these two improvements they study the US Open men's singles 1994–2006, and they find the same as reported in this book: breakpoint does not matter, importance has some (small) impact, and better players perform better than weaker players at important points.

12

Momentum

In both the previous and the current chapters we discuss and test a number of tennis hypotheses, related to the assumption that points are independent and identically distributed. In the previous chapter we concentrated on the identically distributed assumption, now we turn to independence.

Our strategy is the same as in the previous chapter. We first present a simple approach based on averaging, then a more sophisticated approach based on our baseline model. Sometimes the results from the two approaches agree, sometimes they don't. The baseline model provides the more reliable approach, because it properly accounts for quality, importance of a point, and previous-point dynamics. It also produces more reliable estimates of the precisions of our estimators.

Both positive and negative dependence between points in a match should be considered. Winning mood is an example of positive dependence: winning the previous point helps you to win the current point, as we have seen. An example of negative dependence is the break-rebreak effect. A 'break' occurs when a game is not won by the server but by the receiver. The break-rebreak effect is the possible effect that a break increases the probability of a rebreak. Instead of gaining momentum after the break, the current server (receiver in the previous game) is broken back. Does such an effect exist or it just tennis folklore?

Streaks, the hot hand, and winning mood

A big issue in the statistical analysis of sports is whether 'momentum' exists. Momentum can only exist when there is dependence, and dependence can occur between matches and between points within a match.

Consider Esther Vergeer, the Dutch wheelchair tennis player who won 470 matches in a row, being unbeaten from 2003 until she retired in 2013. This is a huge match-winning streak. Her extraordinary talent is one reason for this streak. But perhaps there is an additional (psychological) reason: the more she wins, the lower is the opponent's confidence in beating her. While Vergeer's winning streak relates to dependence between matches, our focus will be on dependence between points within one match.

Because of her talent, Vergeer wins many consecutive points in each match. This is *not* what we mean by momentum. Momentum is measured conditional on both players' qualities, and tells us whether, given her and her opponent's qualities, Vergeer wins more consecutive points than can be explained by randomness. Many will remember the 1993 Wimbledon final between Steffi Graf and Jana Novotna. Novotna led Graf 4-1, 40-30 in the final set with Novotna on service. Instead of winning her service game, she double-faulted and Graf won five consecutive games and the title (one of her twenty-two grand slam victories). This suggests strong dependence between points. One match is, however, not sufficient to prove the existence of dependence in general.

Many studies have addressed momentum in various sports, not only in tennis but also, and in particular, in basketball and baseball. Dependence between points is called a 'streak' in baseball, the 'hot hand' in basketball, and 'winning mood' in tennis. A descriptive study focussing on Detroit's Vinnie Johnson, who has a reputation of being one of the most streaky of all shooters in the National Basketball Association, concluded that he is indeed a streaky shooter and hence that the hot hand exists. But there are other studies that find no dependence. The general evidence from the sports literature, including tennis, is mixed and inconclusive.

People tend to believe in streaks, because casual observation overestimates their occurrence. This is a well-established fact. Suppose, for example, that you receive a letter on Monday predicting next Wednesday's movement of the share prices at the stock exchange (up or down). The prediction is correct. Next Monday you receive another prediction, again correct. This goes on for five weeks. Each time the prediction is correct. Then, on Monday in week six you are asked to send a check for $100 to receive the next prediction. Will you send the check? After all, the writer of the letters was right five consecutive times.

From the point of view of the writer of the letters, the experiment is easy. He or she selects thirty-two people, and writes to each of them: half of them predicting 'up', the other half 'down'. Sixteen people will have received a correct prediction. To these a second letter is sent. Now eight people have received two consecutive correct predictions. After five weeks, one person remains who has received five consecutive correct predictions. No knowledge about the stock market is required. It is pure luck. The probability of five consecutive successes is 1/32 (about 3%), which is small but not as small as the general public feels it to be.

Why study tennis?

Testing whether momentum really exists is difficult. Momentum, if it exists at all, will be small. If the quality of a team (or a player) is underestimated, then there will be more winning streaks by that team than expected by pure randomness. As a result, a small error in measuring quality may lead to the conclusion that momentum exists when in fact it doesn't. The crux of testing for momentum is to have a good correction for quality and good data.

The features of tennis, combined with our Wimbledon data set and our baseline statistical framework, provide the ideal environment for investigating the possible existence of momentum. One important advantage is that in tennis singles matches, only two players meet. There are no player substitutions (in contrast to basketball, say), and this makes it easier to measure quality. Our data are on Wimbledon, where the stakes are high and losing some key points may be decisive for the match outcome. Players have strong incentives to play their best tennis at each point, which again makes it easier to control for quality. We have exploited these facts in Chapter 7 to develop a quality correction that uses observables (world ranking) and accounts for unobservables (form of the day). This correction is included in our baseline model.

A second reason why analyzing tennis can contribute to our understanding of momentum is that a tennis match consists of many points (in contrast to football, say), and the observations are clean (there is virtually no subjective judgement involved). If momentum is small, then a large number of clean observations is required to obtain estimates that are sufficiently precise to provide significant evidence of momentum. Tennis is an ideal setting to test for momentum.

Winning mood in tennis

Momentum in tennis is usually called 'winning mood'. Winning mood can be defined in many ways. We might, for example, ask whether in a final set (fifth set in the men's singles, third set in the women's singles) the player who has won the previous set has the advantage. From our Wimbledon data set, the probability that the same player wins the fourth and fifth sets is estimated to be 50.2%. In the women's singles, the estimated probability that the same player wins the second and third sets is 61.2%. Neither of the two percentages is significantly different from 50%, so there appears to be no basis for this aspect of winning mood.

We shall investigate the possible existence of winning mood, first by considering the previous service point and then by considering the previous *ten* service points of the current server. Our hypothesis is

Hypothesis 21: *Winning mood exists.*

As in the previous chapter, we analyze the hypothesis in two ways: simplistic and not so simplistic.

The simplistic approach is based on Table 12.1. Consider a server in the men's singles. If he won the previous point in his current service game, then the overall probability of winning the current point is 65.3%. This is 2.1%-points higher (significant) than if he lost it. In the women's singles the effect is even higher with a significant difference of 2.8%-points. The existence of winning mood seems clearly visible.

These overall averages, again, ignore quality differences. Having won the previous point may simply indicate that the server is the better player, so that he or she will win the current point even when there is no winning mood. One step towards allowing quality differences is to distinguish between seeded and non-seeded players. The effects are generally smaller now, and we notice that the effect

	Sd-Sd	Sd-NSd	NSd-Sd	NSd-NSd	Total
Men					
Won previous point	66.3	70.0	62.3	64.6	65.3
Lost previous point	67.5	67.8	59.7	62.7	63.2
Difference	−1.2°	2.2*	2.6*	1.9*	2.1*
Women					
Won previous point	56.0	63.8	50.5	57.6	57.4
Lost previous point	56.5	61.6	49.2	54.5	54.6
Difference	−0.5°	2.2*	1.3°	3.1*	2.8*

Table 12.1: *Percentage of points won on service after server won or lost previous point (same game), seeded (Sd) and non-seeded (NSd) players*

is significant in matches between two non-seeded players, but not in matches between two seeded players. This makes sense. If two amateurs meet, and one of them misses a point that he or she should not have missed, then the resulting anger and frustration is likely to have an effect on the next point(s). The better a player is, the faster can he or she concentrate on the current point and forget about the previous point. Table 12.1 confirms this within the group of professional tennis players.

Still, we should be careful in drawing conclusions from this table, because we cannot really disentangle quality and winning mood by just looking at averages of matches. As before, a good quality correction is essential and this is where the baseline model comes in.

The baseline model disentangles quality and the effect of the previous point by including the rankings, unobserved quality, and the dynamic variable $d_{2,it} = f_{i,t-1}$, which equals one (zero) in case of success (failure) at the previous service point. We do not need a new table for this situation, because the estimates in Table 10.3 on page 172 contain all the necessary information.

We find that the previous-point (momentum) effect is significant ($\delta_{20} > 0$) for both men and women. The effect in Table 10.3 is about half the effect in Table 12.1: 0.9 versus 2.1 in the men's singles, and 1.2 versus 2.8 in the women's singles. This is because we now have properly corrected for quality. The incomplete quality correction in the seed/non-seed analysis biases the winning-mood estimate upwardly (sample selection bias), and the baseline model removes this bias. We conclude that our data support the idea of a winning mood. There appears to be a weak positive effect, smaller for top players than for weaker players.

So far we have only looked at the previous point. But this may not be enough. Perhaps a winning mood is based on more points. Let us consider the previous ten points served by the current server. These ten points approximate the service success in the last two service games. This new idea leads to a new dynamic variable,

$$d_{8,it} = \text{relative frequency of winning the previous ten service points.}$$

Obviously there is no value for the first ten service points in a match for each player. For these points we set d_8 equal to the overall average winning probability, as estimated in Table 10.1, that is 64.6% for the men and 55.9% for the women.

	δ_{80} mean	δ_{8-} $r_i - r_j$	δ_{8+} $r_i + r_j$
Men	1.4°	−0.8°	−5.9°
Women	−0.4°	10.4°	9.2°

Table 12.2: *Impact of the previous ten service points*

Our model thus includes both the previous-point dummy d_2 *and* the previous-ten-points dummy d_8 (and of course the importance variable d_1). Table 12.2 shows that all estimates associated with d_8 are insignificant. Adding d_8 to the baseline model is therefore not useful, and our earlier conclusion stands.

Breaks and rebreaks

The server is expected to win his or her service game, particularly on fast surfaces such as grass. If he or she fails to do so (a service break), this is considered serious in the women's singles and very serious in the men's singles. A break could therefore have an effect on players in subsequent points.

In Chapter 5 we briefly discussed hypothesis 8: after a break the probability of being broken back increases. The reason why many commentators believe in this hypothesis is probably that the current server enjoys the success of the break in the previous game and relaxes a bit, while the opponent is eager to strike back. This would induce a negative correlation, in contrast to the positive correlation associated with winning mood.

In Table 5.6 (page 81) we found an unexpected and counterintuitive result. Based on overall averages we concluded not only that the hypothesis should be rejected, but even that there is significant evidence that the opposite is true: after a break, the player is more likely to win his or her own service game. This is contrary to what commentators tell us, and we already remarked that the reason for this result could be that we do not properly account for quality differences.

	Sd-Sd	Sd-NSd	NSd-Sd	NSd-NSd	Total
Men					
After break	66.7	68.9	60.9	65.6	65.8
After no-break	66.7	69.4	61.1	63.4	64.1
Difference	0.0°	−0.5°	−0.2°	2.2*	1.7*
Women					
After break	56.9	62.9	50.4	57.5	57.9
After no-break	56.4	62.9	49.5	54.9	54.9
Difference	0.5°	0.0°	0.9°	2.6°	3.0*

Table 12.3: *Percentage of winning a service point depending on the occurrence of a break in the previous game (same set), seeded (Sd) and non-seeded (NSd) players*

In Table 12.3 we reproduce the overall averages from Table 5.6, and we add a first attempt to allow for quality differences by distinguishing between seeded and non-seeded players. In a match between players of unequal strength (Sd-NSd and NSd-Sd) there is no longer a positive effect. In a match between players of equal strength (Sd-Sd and NSd-NSd) the effect is also insignificant (except NSd-NSd for the men). Controlling for quality differences clearly matters. Maybe a finer quality measure leads to more informative results.

This, again, is where our baseline model enters. We extend the baseline model with one dynamic variable d_9, which should capture the break-rebreak effect. But how should we define this new variable? The simplest definition would be: 1 if there was a break in the previous game (in the same set), and 0 otherwise. This definition would, however, include situations that are not typical for the break-rebreak hypothesis, for example a break at 4-0 or a break at 2-2 following four earlier breaks. What we wish to capture is an *unexpected* break in the previous game, for example at 3-3 or 4-3. After such a break the game score would go to 4-3 or 5-3, a difference of one or two games.

Thus motivated, we capture the break-rebreak effect by the more refined dummy variable

$$d_{9,it} = \begin{cases} 1 & \text{if break occurred in previous game,} \\ & \text{but not in the game before (same set),} \\ & \text{and server leads by one or two games,} \\ 0 & \text{otherwise.} \end{cases}$$

With this definition, the impact of a break-rebreak, while controlling for quality, importance, and previous-point dynamics, is summarized in Table 12.4.

	δ_{90} mean	δ_{9-} $r_i - r_j$	δ_{9+} $r_i + r_j$
Men	0.2°	−2.1°	2.4°
Women	−1.8*	0.9°	1.1°

Table 12.4: *Impact of break-rebreak dummy*

In the men's singles there is no effect, but in the women's singles there is a break-rebreak effect ($\delta_{90} < 0$), and this effect is not rank-dependent. This then is our final conclusion: a break-rebreak effect is present in the women's singles but there is no evidence of the effect in the men's singles.

Our conclusion for this hypothesis changed several times: from a positive effect in Chapter 5 using overall averages, via a zero impact using a quality correction that distinguishes only between seeds and non-seeds, to a negative effect (for the women) when we control appropriately for quality. The counterintuitive results from the simplistic approaches do not survive a more serious statistical analysis.

Missed breakpoints

In the Federer-Nadal 2008 Wimbledon final, Roger Federer was 4-3 ahead in the final set and had breakpoint on Rafael Nadal's service, but he missed it and Nadal held serve. Two games later at 5-4, Federer needed only two points (on Nadal's serve) to win the championship. Nadal held serve, and at 5-5 Federer was two breakpoints down. Commentators were quick to point out that Federer's disappointment of missing the earlier chances had affected the momentum in Nadal's favor (page 38). Were the commentators right or wrong in drawing this conclusion?

To find out, we consider the situation where in the previous game no break has occurred, although the receiver had a good chance to break because he or she had one or more breakpoints. The receiver did not capitalize on these breakpoints and this may have discouraged him or her. Does such a 'discouragement effect' occur? And, if so, does it influence the performance in the current game where the discouraged receiver from the previous game is serving?

Hypothesis 22: *After missing breakpoint(s) there is an increased probability of being broken in the next game.*

Table 12.5 shows no compelling support for this idea. For the men, the probability of winning a point on service after missing all breakpoints is 0.4%-points lower than usual (1.7%-points for the women), in line with the hypothesis but not significant. When we bring in quality and distinguish between seeds and non-seeds, the differences remain largely insignificant. A simple analysis therefore does not support this hypothesis.

	Sd-Sd	Sd-NSd	NSd-Sd	NSd-NSd	Total
Men					
Missed breakpoints	66.6	70.3	60.2	62.5	63.8
No breakpoints	66.8	69.2	61.2	63.6	64.2
Difference	−0.2°	1.1°	−1.0°	−1.1°	−0.4°
Women					
Missed breakpoints	55.5	62.1	50.1	50.9	53.5
No breakpoints	56.6	63.1	49.4	55.6	55.2
Difference	−1.1°	−1.0°	0.7°	−4.7*	−1.7°

Table 12.5: *Percentage of winning a service point depending on the occurrence of missed breakpoints in the previous game (same set), seeded (Sd) and non-seeded (NSd) players*

The conclusions are different when we employ our baseline model, where we now add the dynamic dummy

$$d_{10,it} = \begin{cases} 1 & \text{if breakpoint(s), but no break,} \\ & \text{occurred in the previous game (same set),} \\ 0 & \text{otherwise.} \end{cases}$$

The results are given in Table 12.6. While the evidence is not significant in the men's singles, there is evidence for the hypothesis in the women's singles: discouragement because of missed breakpoints in the previous game leads to a smaller probability of winning a point in the current game ($\delta_{10,0} < 0$).

Moreover, the fact that $\delta_{10-} < 0$ for the women indicates that missing breakpoints against a weaker opponent is particularly harmful. This is consistent with the discouragement idea: missing breakpoints that a player should not have missed, because she is better than her opponent, is particularly disappointing.

Finally, $\delta_{10+} > 0$. This means that if we compare two matches with the same quality differential $r_i - r_j$, then the better of the two matches (highest $r_i + r_j$) exhibits the least impact of missed breakpoints. In fact, in a top match where r_i and r_j are equal and the (centered) $r_i + r_j = 0.8$, the impact of missed breakpoints is virtually zero: $-3.8+5.4\times 0.8 = 0.5$. This adds support to the idea

	$\delta_{10,0}$ mean	δ_{10-} $r_i - r_j$	δ_{10+} $r_i + r_j$
Men	-1.2°	1.6°	-0.6°
Women	-3.8^*	-7.8^*	5.4^*

Table 12.6: *Impact of missed breakpoints*

that top players are not only technically but also mentally stronger than other players, thus reinforcing our findings of Chapter 10. Top players do not let themselves be discouraged and are less affected by what happened in the past.

The encompassing model

The restriction $d_{10} = 0$ considers two cases jointly: either no breakpoints occurred in the previous game or a break did occur. It makes sense to combine d_{10} with d_9 so that the two cases can be considered separately. Thus we also estimate the baseline model with d_9 and d_{10} added jointly rather than separately, but the conclusions remain the same.

Such analyses provide a check on the robustness of the estimation results. This is important, because estimates depend on the assumptions underlying the model, and if those assumptions are violated inference may become unreliable. The simplistic analyses presented in this book and the improved results from applying better methods provide numerous examples of this fact.

Statistical studies typically contain many robustness checks, and we performed numerous checks to verify that the models used in this book are robust, that is, not sensitive to small changes in the specification of the deterministic and random components of the model. We do not elaborate on these checks, except in the current section where we present the estimation results when *all* dynamic variables are added to the baseline model. Thus, in Table 12.7, we add to importance (d_1) and previous point (d_2) the dynamic variables d_3 to d_{10} jointly, not separately.

When we compare Table 12.7 with the eight tables in Chapters 10 to 12 based on including each variable separately, we find remarkable robustness. The results change, but they never change

	Men			Women		
	δ_0 mean	δ_- $r_i - r_j$	δ_+ $r_i + r_j$	δ_0 mean	δ_- $r_i - r_j$	δ_+ $r_i + r_j$
Importance	−8.2°	32.4°	42.7*	−9.2°	121.4*	16.6°
Previous point	0.7°	0.3°	−2.2°	0.9°	2.3°	−4.4*
New balls	1.4°	−3.2°	−3.4°	−0.5°	0.6°	2.6°
Age of balls	0.2°	0.2°	−0.5°	−0.1°	−0.6°	−0.5°
Breakpoint	−1.1°	1.8°	−3.2°	−2.4*	−4.5°	0.4°
Started serving set	−0.7°	−0.5°	1.8°	0.0°	2.2°	−6.0*
First game in match	3.9*	0.6°	−5.2°	3.1*	−10.6*	−0.3°
Previous ten points	1.1°	0.6°	−7.1°	−1.5°	11.4°	9.5°
Break-rebreak	0.1°	−2.0°	2.6°	−2.1*	−0.5°	1.0°
Missed breakpoints	−1.2°	1.2°	−0.3°	−4.0*	−8.4*	5.0°

Table 12.7: *The encompassing model: impact of variables on the probability of winning a service point*

so much as to lead to different conclusions. The robustness (lack of sensitivity) is important for the validity of our results, equally important as the statistical significance of our results.

The power of statistics

Sometimes the use of an advanced model makes a difference, sometimes it doesn't. With some of the hypotheses that we have discussed, a simple analysis based on averages provides the same conclusion as a more sophisticated analysis that appropriately takes account of quality and other key variables. But with other hypotheses the conclusion changes, and the more sophisticated approach is essential. We do not know in advance how complex we should make our model.

In Chapter 2 we quoted Albert Einstein, who said: 'As simple as possible, but not simpler'. This is very true and very useful as a basic research philosophy, but it is not easy. The only way to find out whether an approach is too simple or not is to add 'credible complications'. In this book, these credible complications are the GMM approach and the resulting baseline model. On the whole it appears that simple averaging is too simple, and that the baseline model and the statistical machinery required to estimate it are really necessary in order to obtain significant, credible, and robust results. Thus, we witness the power of statistics in action.

Further reading

The existence of momentum has been studied in several sports, including tennis, and the relevant literature has been reviewed by Bar-Eli *et al.* (2006) and Reifman (2012). Both reviews find very little evidence of the existence of momentum. Athletic streaks, for example, occur about as often as one would predict by random chance. The notion of momentum in sports appears to be greatly overstated. Despite the limited evidence, many people believe in momentum. Amos Tversky attempted to fight this common belief but without success. 'I've been in a thousand arguments over this topic, won them all, but convinced no one', he used to say (quoted in Bar-Eli *et al.*, 2006).

Our analysis suggests that we should distinguish between top players and weaker professionals. We find no momentum for the top players, but we do find some limited momentum for the weaker players. Since there are only a few top players and many somewhat weaker players, many matches will have (small) momentum swings.

Dependence between points in baseball, if it exists, is called a 'streak'. Lindsey (1961) concluded that scores can be well replicated by assuming independence of runs scored in different innings. Siwoff *et al.* (1987) found that the probability of hitting well in a game is independent of whether or not the hitter is 'on a streak'. Albright (1993) agrees and finds no convincing evidence of streaks either. However, in a comment on Albright's article, Stern and Morris (1993) and Albert (1993) suggested alternative approaches that might lead to a different conclusion, thus challenging the independence assumption.

In basketball, dependence between points is known as the 'hot hand'. Research on the hot hand originated with Gilovich *et al.* (1985), who concluded that people believe in the hot hand (not that a hot hand actually exists). Their results are summarized in Kahneman (2011, pp. 116–117) who writes: 'The hot hand is entirely in the eye of the beholders, who are constantly too quick to perceive order and causality in randomness. The hot hand is a massive and widespread cognitive illusion'. Larkey *et al.* (1989) examined game data for eighteen players from the National Basketball Association, in particular Detroit's Vinnie Johnson. From their descriptive analysis they concluded that the hot hand exists.

In tennis, one may ask whether a back-to-the-wall effect exists, where the player who is behind performs better, an illustration of negative dependence: failure breeds success. Croucher (1981) found only very slight evidence of this effect. Jackson and Mosurski (1997) investigated whether 'getting slammed during your first set might affect your next'. In other words, they challenged the independence assumption and concluded that there is positive dependence — failure breeds failure — possibly caused by 'psychological momentum'.

Typical analyses of dependence study how winning or losing past points or sets affects current performance. In contrast, the break-rebreak and missed breakpoint hypotheses 8 and 22 are concerned with specific features of those past points. Tests of these hypotheses (in a less sophisticated framework) were first discussed in Magnus and Klaassen (1996, 1997). These two papers also contain tests of related hypotheses not covered in this book.

13

The hypotheses revisited

Through this book, like a continuous thread, run twenty-two hypotheses. These hypotheses are statements on which tennis gurus hold strong views: serving first in a set is an advantage, winning mood exists, serving with new balls is an advantage. Of course we do not accept the tennis gurus' wisdoms without solid statistical evidence, and therefore we test the hypotheses using statistical techniques, sometimes simple, sometimes more complex. Simple techniques are sometimes sufficient (which we typically don't know until we perform a more sophisticated test), sometimes not. Some hypotheses turn out to be true, some are false.

In this final chapter we summarize what we have found. Our aim is to offer a clear answer, for each of the twenty-two hypotheses, to the question of whether the hypothesis is true or false. The book contains more than the hypotheses (for example, a discussion of how to forecast the winner of a match and how to choose an optimal service strategy), but in this chapter we focus on the hypotheses only. The hypotheses are presented in the same order as they appear in the book.

In formulating our answers in this chapter, we put clarity above statistical cautiousness and subtlety. For example, instead of saying that a hypothesis is rejected or not rejected (as statisticians like to do), we shall now simply say that we reject or accept the hypothesis.

1 Winning a point on service is an iid process

Our first hypothesis, not the easiest, is important for our theoretical framework, and its outcome matters because later hypotheses (from 15 onwards) only make sense when this first hypothesis is rejected.

What is iid? It means: independent and identically distributed. When points are *dependent*, then what happened at the previous point influences the current point. A missed smash may negatively affect the player's next point, and a series of good points may create a winning mood and affect the next point positively. When points are *not identically distributed*, then a player will be affected by the score and possibly play more cautiously at important points such as a breakpoint.

Considering the service points of one player in a singles match, we can ask two questions: are these points independent or not? and are they identically distributed or not? When both effects are absent, then the service points are independent and identically distributed.

If the hypothesis were true, then there would exist no winning mood, no break-rebreak affect (two examples of dependence), and no cautious play at important points (non-identical distribution). Players then play each point 'as it comes', without being affected by what happened at previous points or by the score. Few readers will believe this and they are right: we find that service points are neither independent nor identically distributed. What is remarkable though, is not so much this conclusion (which we expected), but rather the fact that the deviation from iid is small.

This immediately raises a further question: is this rejection of iid true for all players or is there perhaps a difference between top and other players? We shall see later (in hypothesis 17) that there is indeed a difference, apart from the obvious difference in playing strength, between top players and other players.

2 It is an advantage to serve first in a set

This hypothesis, our favorite, has been discussed throughout the book. When we consider individual sets and ask whether the player who served first also wins the set, then we find that this is possibly the case in the first set, but not in any other set. On the contrary, in the other sets serving first is a *disadvantage*. This is an unexpected result and it holds for both men and women.

The result is unexpected, but it can be explained as follows.

The player who serves first in a set, if it is not the first set, is usually the weaker player. This is so because of a combination of two factors. First, it is likely that the stronger player won the previous set; second, it is likely that the last game of the previous set was won by the server. As a result, the loser of the previous set typically serves first in the next set. He or she is then more likely to lose the current set, not because of a first service (dis)advantage, but because he or she is the weaker player.

It is true therefore that serving first is *correlated* with losing the set, but it is not true that serving first *causes* losing the set. The cause is that the weaker player typically serves first.

To dig deeper we need to adjust for the fact that it is the weaker player who typically serves first in a set (if it is not the first set). In other words, we need to take the quality of the players into account. If we do take quality into account (their rankings, form of the day, and so on), then we find that there is no advantage (and no disadvantage) in serving first in a set. So, the hypothesis is rejected, for both men and women. This holds for the second and subsequent sets. Whether it also holds for the first set remains to be investigated, and we shall do so in hypothesis 20.

3 Every point (game, set) is equally important to both players

One often hears: 'This is an important point, especially for so-and-so', where so-and-so is typically the player who is behind. If he or she is down 15-40 in a game, is it perhaps more important for the player who is behind to save this breakpoint than for the player who is ahead to convert it? No, it is not. Each point is equally important to both players.

No statistical analysis is provided for this hypothesis. It is not a question of statistics, but of logic. If there are only two players, then an advantage for one is automatically a disadvantage for the other: if saving a breakpoint increases the winning probability of the server by 1%, then it automatically increases the losing probability of the receiver, also by 1%.

This does not mean that each point is equally important (hypothesis 5): some points are more important than other points. It means that if one point is important to the server it is also important to the receiver. If another point is more important, then it is more important to *both* server and receiver.

4 The seventh game is the most important game in the set

The sonorous tones of Dan Maskell were, for several decades, as much a part of Wimbledon as the Duchess of Kent and strawberries and cream. When commenting on a particularly exciting piece of play or an outstanding shot, he sometimes used his most remembered and revered catchphrase 'Oh, I say', which might well be all he would say during an entire set.

Maskell was strangely fixated on the seventh game of the set, which he never failed to label 'all-important'. One way to approach this hypothesis is to argue that not the seventh game (or indeed any other game) but the tiebreak is the most important game in a set. Winning or losing the seventh game may be important, but surely winning or losing the tiebreak is more important. The hypothesis is therefore rejected.

This analysis is correct, but a tiebreak can only be important when there is a tiebreak. Most sets, however, don't reach 6-6 and therefore don't have a tiebreak. A more informative analysis needs to take into account that some games occur less frequently than others. If we do that, thus accounting for the fact that the higher game numbers occur less frequently, then we find nothing special about the seventh game. So, also from this viewpoint, the hypothesis is rejected.

5 All points are equally important

Certainly not. In a game, 0-0 is less important than 30-40. In a set, 0-0 is less important than a tiebreak. In a match, the final set is more important than the first.

Matchpoint is usually not the most important point in a match. For example, at 2-0, 5-0, 40-0, matchpoint is not important at all. The most important point is the point where the swing in the match-winning probability between winning the point and losing it is largest. Points at the end of a tiebreak and breakpoints typically have this property.

6 The probability that the service is in is the same in the men's singles as in the women's singles

Remarkably, the percentage of first and second service in is almost the same for men and women. On average, the first service is in 59.4% of the time for the men and 60.8% for the women, while the second service is in 86.4% of the time for men and 86.0% for women. The difference between men and women is small (1.4%-point) for the first service and even smaller (0.4%-point) for the second service. These percentages are approximations (estimates) of the underlying probabilities. The probabilities themselves cannot be observed — only the percentages of services in. Can we conclude from these percentages that the probabilities are the same for men and women?

The degree of complexity of our statistical analysis matters here, and our conclusion differs between first and second service. For the first service we conclude that the probability that the service is in is *not* the same in the men's singles as in the women's singles, in spite of the fact that 59.4% and 60.8% are close. The percentages are close but not close enough from a statistical perspective. But for the second service we conclude that the probability that the service is in is the same.

7 The probability of a double fault is the same in the men's singles as in the women's singles

Closely related to the previous hypothesis is this statement about double faults. For the average player we predict 5.52% double faults for the men and 5.49% for the women at Wimbledon. These percentages are very close. How close are the underlying probabilities? Is the probability of a double fault the same in the men's singles as in the women's singles? When we test this hypothesis, we accept it. Hence, men and women have the same probability of serving a double fault.

How do we reconcile this conclusion with the conclusion from the previous hypothesis, in particular the fact that the probability of first service in is not the same for men and women? One reason is estimation uncertainty. A double fault occurs when both the first service and the second service are a fault. As a result, the probability of a double fault depends on both the first-service-in and the second-service-in probabilities. In the previous hypothesis we accepted one thing (second-service-in probability is the same for men and women), while we rejected the other (first-service-in probability is not the same for men and women). Whether or not we reject the combination (same probability of a double fault) depends on the estimation uncertainty associated with both the first and the second service. In this case, the combined estimation uncertainty is so large that we cannot reject the combined hypothesis. Thus we conclude that men and women have the same probability of serving a double fault.

8 After a break the probability of being broken back increases

Unexpected conclusions can emerge when we do not analyze the data in a statistically sound manner. We saw this throughout the book, for example in hypothesis 2, where we asked whether it is an advantage to serve first in a set. The famous break-rebreak hypothesis provides another example.

If the hypothesis is true, then the probability of winning a service game decreases after a break of the opponent's service in the previous game (in the same set), so that a rebreak becomes more probable. A simple analysis reveals that the probability of winning a service game is 3.3%-points *higher* if a break occurred in the previous game than if no break occurred. This is for the men; for the women the difference is 5.7%-points, even higher. Apparently, after a break the probability of being broken back decreases rather than increases.

Of course it does, if we don't take quality into account! The better player breaks the opponent's service *and* holds his or her own service, not because a break makes it easier to hold service, but because he or she is the better player.

When we correct for quality, then we find no effect in the men's singles but we do find an effect in the women's singles: women who have broken their opponent's service should be extra careful not to lose their own service game. So, we reject the hypothesis for men but accept it for women.

9 Summary statistics give a precise impression of a player's performance

Our data set is large and clean, so one would expect that summary statistics involving all matches are informative. This is indeed the case. But let us consider just one match, say the Djokovic-Nadal Australian Open 2012 final. The viewer is presented after each set with summary statistics of that set and of the match. How informative are these statistics?

In the Djokovic-Nadal match, the first two sets were long and close. Djokovic lost the first set 5-7 and won the second 6-4. The third set was short and Djokovic won it 6-2, suggesting a dip in Nadal's performance. But was this dip structural or was it just chance?

Nadal's first-service percentage dropped from 71% in the first two sets to 53% in the third set. It would be tempting to conclude that Nadal took more risks on his first service in the third set, and that it would be better to take less risks to get the percentage up again in the fourth set. However, realizing that the percentages are based on ninety-two points (in the first two sets) and only thirty (in the third set), the statistically correct conclusion is that there is no evidence that Nadal changed anything. The difference in percentages can easily be explained by chance. Hence, to conclude from these percentages that Nadal changed his service strategy in the third set is not justified from a statistical viewpoint.

This does not mean that summary statistics are useless. They reflect what actually happened and they are entertaining. However, due to the presence of chance, they do not equal but only approximate a player's underlying performance, and the smaller the number of points on which the summary statistic is based, the larger is the relevance of chance, and the poorer is the approximation. In general, summary statistics on individual matches do not give a precise impression of a player's performance.

10 Quality is a pyramid

The 'quality' of a player is obviously important in explaining his or her success. How to take quality into account requires careful modeling, as we have seen in hypotheses 2 (advantage of serving first in a set) and 8 (break-rebreak effect), and failure to do so can lead to unexpected results. Quality is partly observed (through the ranking) and partly unobserved (form of the day, small injuries, fear of a specific opponent). Unobserved quality can be accounted for, even in the absence of data, and we elaborate on this subtle issue in this book. The question underlying the current hypothesis, however, deals with observed rather than with unobserved quality: how do we measure observed quality?

The data consist of the ranking points for each player, published weekly by the ATP (for men) and WTA (for women). The 2012 end-of-year lists show Novak Djokovic (12,920 points) and Victoria Azarenka (10,595 points) as the world's number one at the end of 2012. The same two lists show that there are five players in the men's singles (three in the women's singles) who have between 2000 and 2500 points, seven (thirteen) between 1500 and 2000 points, seventeen (thirty-two) between 1000 and 1500 points, and seventy-three (seventy-eight) between 500 and 1000 points. So, going down from the top by steps of 500 points each involves more and more players. This is what we mean by a pyramid, and a more detailed analysis confirms that quality is indeed a pyramid.

This also helps in answering the next question: how do we measure quality? The ranking points lead to a ranking $(1, 2, \dots)$ and we could say: quality equals ranking. This would *not* be a good definition, however, because it would imply that the difference in strength between, say, numbers 1 and 16 is the same as the difference between 101 and 116, and we know now that this is not the case.

We could also say: quality equals ranking points. This would lead to a pyramid, but not necessarily the best pyramid. A problem with this definition is also that the method of computing ranking points changes regularly and is not the same for men and women. In this book we developed our own formula for quality, obviously a pyramid based on the rankings.

Our formula is based on the idea of 'expected round'. The expected round is a number associated with the ranking: 8 for a player with $rank_i = 1$ who is expected to reach the final (round 7) and win, 7 for a player with $rank_i = 2$ who is expected to reach the final and lose, 6 for players with $rank_i$ 3 or 4 who are expected to lose in round 6, 5 for players with $rank_i$ 5 to 8 who are expected to lose in round 5, and so on. When we walk down the pyramid, more and more players get the same quality indicator, capturing the flattening of quality differences represented by the idea of a pyramid. A problem with the expected round, however, is that it does not distinguish between, for example, players ranked 9 to 16, because all of them are expected to lose in round 4. Our formula — a smoothed version of expected round — provides expected-round numbers for all rankings, is a pyramid, and captures the essence of quality.

11 Top players must grow into the tournament

For a top player in an early round, the actual round will be smaller than the expected round based on his or her ranking. Does the top player perform according to his or her expected round or somewhat less good, because he or she has to grow into the tournament? There is no evidence of this in our data set. Professional tennis, especially at grand slam tournaments, is so competitive that top players can not and do not relax in the early rounds.

12 Men's tennis is more competitive than women's tennis

This is true. Since men play the best-of-five sets and women the best-of-three, one would expect fewer upsets in the men's singles than in the women's singles: the longer the match format, the more likely it is that the favorite will win. However, when we consider the data, we see that *more* upsets (top-sixteen seeds not reaching the last sixteen) occur in the men's singles than in the women's singles. Men's tennis is therefore more competitive than women's tennis, and a more sophisticated statistical analysis confirms this conclusion.

In recent years, women's tennis has become more competitive. The world's number-one position has changed regularly and many different women have won one or more of the grand slam tournaments. In contrast, men's tennis has been dominated by just four players: Novak Djokovic, Roger Federer, Rafael Nadal, and Andy Murray. Casual observation may suggest that men's tennis is no longer more competitive than women's tennis. This suggestion is not supported by the data. While it is true that the competitive gap between men's and women's tennis has become smaller, it is still the case that men's tennis is more competitive than women's tennis.

13 A player is as good as his or her second service

To investigate this often-heard hypothesis we first need to define when a second service is 'good'. A (first or second) service is 'good' when the percentage of points won on that service is high. Remarkably, this characteristic is typically not shown on television. The characteristics shown are the percentages '1st (2nd) service in' and 'points won if 1st (2nd) service is in'. But it is not these two probabilities themselves which matter most — it is their product. Hence, a second service is good if the product of the percentages '2nd service in' and 'points won if 2nd service is in' is high.

Next we need to define what we mean by a 'good' player, and determine what type of matches are therefore appropriate to test the hypothesis. We could compare matches between a seeded server against any opponent with matches between a non-seeded server against any opponent. The server in the first type of match is a better player than the server in the second type, and this allows us to test the hypothesis by checking whether the seeded server wins more points on second service than on first service compared to the non-seeded server.

The previous comparison distinguishes between the quality of servers, but it does not distinguish between the quality of receivers. If we wish to allow for this distinction as well, we should compare matches between a seeded server against a seeded receiver with matches between a non-seeded server against a non-seeded opponent. As a third possibility we could compare matches between a seeded server against a non-seeded opponent with matches between a non-seeded server against a seeded opponent.

The three types of matches correspond with different interpretations of the hypothesis. We analyze each type separately, not only using the rough distinction seeded versus non-seeded, but in particular using a finer distinction based on our definition of quality. We conclude that a player is indeed as good as his or her second service.

14 Players have an efficient service strategy

Each server in a tennis match chooses that strategy which maximizes the probability of winning a point. A good strategy involves making both services neither too easy (in which case the receiver will kill it) nor too difficult (in which case the service will too often be a fault). In this book we developed a model to answer the question how difficult a player should make his or her service in order to maximize the probability of winning a service point. With this model we can compute the optimal service strategy and compare it to the actual strategy. The ratio between the actual probability and the optimal probability is the 'efficiency', a number between zero and one. If the ratio equals one, then there is full efficiency. Of course, the efficiency differs per player and depends also on the opponent.

For a player in a given match, the efficiency could be one, but more likely it will be lower than one. How much lower? We find that the average efficiency is 98.9% for the men and 98.0% for the women. This is pretty efficient, but not fully efficient, and we reject the hypothesis.

We reject the hypothesis, but our method also allows us to quantify the inefficiency: 1.1% for men and 2.0% for women. Apparently, the women are less efficient than the men, at least in choosing their service strategy. We also find that 25% of the men have an inefficiency of more than 1.4% and that 25% of the women have an inefficiency of more than 2.8%.

Inefficiencies thus vary substantially across players. This variation is not pure chance; there is structure in it. Higher-ranked players are more efficient than lower-ranked players: they are not only better tennis players, but they also make better choices about what type of service to use.

15 Players play safer at important points

In hypothesis 1 we saw that it is not the case that players play 'one point at a time', that is, we rejected the hypothesis that points are iid. In fact, service points are both dependent *and* non-identically distributed, although the deviations from iid are small. In the remaining hypotheses of this chapter we try to discover more about these deviations.

Players may be tempted to serve with less risks at important points, in order to make sure that the first service is in and to avoid a double fault. And indeed we find, at important points, that more first services go in, and fewer aces and double faults occur. Players serve more conservatively under pressure: they play safer at important points.

However, not all players are the same. Maybe some players are not affected by important points. This key question is discussed below in hypothesis 17.

16 Players take more risks when they are in a winning mood

While the previous hypothesis provided an example of how players deviate from the assumption that points are identically distributed, the current hypothesis asks why players deviate from the assumption that points are independent. Players do indeed take more risks when they are in a winning mood. Winning the previous point reduces the probability of hitting the first service in, and it increases the probability of an ace, but also of a double fault.

Here, too, we may ask whether this result holds for all players or whether we can distinguish between different types of players. Maybe some players are not affected by a winning mood. This question is taken up in the next hypothesis.

17 Top players are more stable than others

Players are trained by their coaches to be mentally stable, forget about the score, forget about the past, and focus on the current point. In other words, they are trained to play points as iid-like as possible. But maybe not all professionals are equally successful in putting this advice into practice. This is true. Top players play like iid, but the lower a player is on the rankings the further he or she deviates from iid. Top players are indeed more stable.

Let us investigate both aspects of iid, first the identical distribution aspect, then the independence. If a top player serves at an important point against a weaker opponent, he or she wins more often than a weaker player would, not only because the stronger player wins more points anyway, but also because he or she is more successful at important points. This is *not* because the top player raises his or her level, but because the weaker player underperforms. Top players are not affected by the importance of a point. They play as if they are ignorant about the score. For top players, service points are identically distributed.

Next, the independence aspect. For a weak (not top) server, winning the previous point has a positive impact on the current point, and losing it has a negative impact. This is the case for both men and women. But top players play the points independently. They are able to switch off the impact from the previous point and 'forget about the past'. For top players, service points are independent.

In contrast, weaker players deviate from iid in two ways: their performance at the current point depends on what happened at the previous point (not independent) and they are also affected by the importance of a point (not identically distributed).

Top players are, of course, better players than weaker players. But they have, in addition, two advantages that are unrelated to their service power and technique. They are more stable than weaker players and, as we have seen in hypothesis 14, they also have a more efficient service strategy.

18 New balls are an advantage to the server

At grand slam tournaments, six new balls are provided at the beginning of the warm-up, then after the first seven games, and then after every nine games. It may seem unlikely that the characteristics of the balls change in such a short lifespan. Still, this is the case.

New balls, just out of the can, are smooth and bouncy, whereas used balls are softer and fluffier. The used balls provide more grip, making it easier to control the service. More first services will go in and fewer double faults occur. On the other hand, the smooth new balls travel faster through the air, so that if the first service is in, the server will have a higher chance of winning the point.

Thus we have two forces working against each other: the probability that the service is in decreases when balls are new, but if the service is in, the probability of winning the point increases. These two forces turn out to possess about the same strength, so that they cancel each other out and no new-ball advantage can be found from the data.

19 Real champions win the big points

Closely related to hypothesis 17 is the current hypothesis, which relates to the identical distribution of service points. We already know that points are not identically distributed. We also know (hypothesis 15) that most players play safer at big points, that is, at points that are especially important, such as breakpoints and points at the end of a close set.

Is it true that top players perform particularly well on those points? Yes and no. The top players are more successful than weaker players, but not because they play better. What happens is that their opponents play worse. This gives the top players a double advantage. Not only are they better players, but they are also more stable under pressure (as we have seen in hypothesis 17), while the opposition underperforms when it really matters.

20 The winner of the toss should elect to serve

In hypothesis 2 we saw that there is no advantage in serving first in a set, except possibly in the first set. We find that the winner of the toss, if he or she elects to serve, has a 3- to 4%-points higher probability of winning a point in the first game, irrespective of his or her quality. This effect exists only in the first game, not in the first two games as one might think. Hence, in general, the winner of the toss should elect to serve.

21 Winning mood exists

The final two hypotheses are concerned with (in)dependence. In hypothesis 16 we found that players take more risks when they are in a winning mood, that is, after winning the previous service point. But what exactly is a winning mood (also known as 'momentum') and does it actually exist? This is a big issue in sport statistics, not just in tennis but in all sports, and it has been studied extensively, particularly in baseball (streaks) and basketball (hot hand). All commentators and almost all spectators believe in it, but that does not mean that it actually exists.

If we define winning mood by only considering the previous service point, then we find evidence of a winning mood because of a previous-point effect, both for men and for women. Our data thus support the idea of a winning mood. The effect is small, much smaller than most spectators believe it to be. When we distinguish between top players and somewhat weaker players, then no winning-mood effect for top players and a small but nontrivial effect for somewhat weaker players can be detected.

One might object that there is more to winning mood than just winning the previous point. This is a reasonable objection, but if we consider not one but ten previous service points (about two service games), then we reach the same conclusion.

22 After missing breakpoint(s) there is an increased probability of being broken in the next game

In hypothesis 8 we studied the impact of a break in the previous game on the performance of the server in the current game. Suppose now that no break occurred in the previous game, but that the receiver had a good chance to break because he or she had one or more breakpoints. The receiver did not, however, capitalize on these breakpoints and this may have discouraged him or her. Does such a 'discouragement effect' occur? And, if so, does it affect the current game where the discouraged receiver from the previous game is serving?

There is no support for this hypothesis in the men's singles, but there is support in the women's singles. Moreover, in the women's singles, the weaker the opponent, the larger is the discouragement effect of missing breakpoints.

This discouragement effect is smaller at high-level matches (say, between two top players) than at lower-level matches. Top players, women as well as men, are not only technically but also mentally stronger than other players. They do not let themselves be discouraged and are less affected by what happened in the past.

A

Tennis rules and terms

In this appendix we briefly summarize the rules of tennis and provide a glossary of some tennis terms used in this book that may not be familiar to all readers. The International Tennis Federation (ITF) is the governing body of tennis and determines the Rules of Tennis. The latest Rules can be downloaded from the ITF website (www.itftennis.com). A more elaborate overview of tennis terminology is available on many websites.

Tennis rules

A match consists of sets, a set consists of games, and a game consists of points. A singles match is contested between two players. The server of the first game is decided by a coin toss. Thereafter, each successive game is served by the other player.

At the beginning of each point one player, the server, hits the ball to the opponent, the receiver. When the service is a fault, the server hits a second service. When the second service is also a fault, the server loses the point. When the first or second service is in, the rally begins. The rally ends when a player fails to hit the ball into the correct court before it bounces twice consecutively, in which case that player loses the point.

A game is finished when one player wins at least four points with a difference of at least two points. The points are counted as 0 ('love'), 15, 30, 40 (rather than as 0, 1, 2, 3). When the score is a tie at 40-all, it is called 'deuce'. When the server wins (or loses) the point after deuce, it is advantage to the server (or receiver). When the player with advantage wins the next point he/she wins the game; otherwise the score is deuce again.

A tiebreak is a special type of game. It is won when a player reaches at least seven points and has a lead of two or more points. In contrast to standard games, points in a tiebreak are counted 0, 1, 2, ... The player who received in the preceding game serves first in the tiebreak. After the first point the service alternates every two points.

A set is finished when one player wins at least six games with a difference of at least two games. When the score reaches 6-6, either the set continues until a two-game difference is achieved or it is decided by a tiebreak, depending on tournament rules. In the latter case the possible winning scores in a set are: 6-0, 6-1, 6-2, 6-3, 6-4, 7-5, or 7-6.

A match consists of a maximum of three or five sets. Women's matches are always best-of-three sets whereas men's matches are either best-of-three or best-of-five, depending on tournament rules.

Tennis terms

Ace: Service where the ball is served in and not touched by the receiver.

Ad court: Left side of the court of each player, so called because the 'ad' ('advantage') point is served to this side of the court. See also *deuce court.*

Advantage: Score in a game where a player has won the point at deuce and therefore needs one more point to win the game.

Break (of service): Game won by the receiving player.

Break back: See *rebreak.*

Breakpoint: Point which, if won by the receiver, results in a break of service.

Challenge: Protest where a player requests an official review of the spot where the ball has landed, using electronic ball-tracking technology; see *Hawk-Eye.* Challenges are only available in some tournaments and at some courts.

Deuce: Score of 40-40 in a game.

Deuce court: Right side of the court of each player, so called because at deuce the point is served to this side of the court.

Double fault: Two service faults in a row within one point, causing the player serving to lose the point.

Doubles: Matches played by four players, two on each side.

Final set: Third set in a best-of-three match; fifth set in a best-of-five match.

Gamepoint: Score where the server needs one more point to win the game. See also *breakpoint*.

Grand slam: The grand slam tournaments are the four major tournaments in a calendar year: the Australian Open, the French Open (or Roland Garros), Wimbledon, and the US Open.

Hawk-Eye: Computer system connected to cameras to track the path of the ball for replay purposes.

Hold (service): To win the game when serving. Compare *break*.

In: A ball is in if, after leaving the racket, it first hits the correct court. For example, a service is in if the ball first hits the service court diagonally opposite.

Matchpoint: Score where a player needs one more point to win the match.

Rally: Following the service, a series of hits that ends when the point is decided.

Ranking (world ranking): Official ordering of players, updated weekly, based on the number of points he or she accrued during the past year. The best player is ranked number one.

Rebreak: To win a game as the receiving player immediately after losing the previous game as the serving player.

Receiver: Player who is being served to.

Seed: Highly-ranked player whose position in a tournament has been arranged based on his/her ranking so as not to meet other highly-ranked players in the early rounds of play. For a given tournament there is a specified number of seeds, depending on the size of the draw. For ATP/WTA tournaments, typically one out of four players are seeds.

Server: Player who is serving.

Service (serve): Every point begins with a serve where the server attempts to hit the ball over the net into the service court diagonally opposite.

Service court: On each side of the net there are two service courts: left and right. These are boxes bounded by the singles sidelines, the serviceline, and the center serviceline.

Service game: With regard to a player, the game in which the player is serving (e.g., 'player $\mathcal{I}$ won a service game on love' means that $\mathcal{I}$ won a game where (s)he was serving without the opponent scoring a point).

Setpoint: Score where a player needs one more point to win a set.

Singles: Matches played by two players, one on each side.

Tiebreak: Special game played when the score is 6-6 in a set to decide the winner of the set.

Toss: Before the beginning of a match a coin is tossed, usually by the umpire. The toss decides the choice of ends and the choice to be server or receiver in the first game.

Unseeded player: Player who is not a seed in a tournament.

Winning mood: Mental condition of a player after being successful in the preceding points or games, possibly leading to a higher probability of being successful at the current point.

B List of symbols

In this second appendix we list all symbols used in the book, ordered by type. We have aimed for consistency in notation, realizing that complete consistency, even if possible, is not desirable because it comes at the cost of too many qualifiers.

In a typical tennis match, two players $\mathcal{I}$ and $\mathcal{J}$ are competing against each other. Many variables thus depend on i (corresponding to $\mathcal{I}$) and j (corresponding to $\mathcal{J}$). For notational simplicity we typically leave out the j index. The index t for the service point number is also often left out, particularly when a variable is constant over time.

Winning probabilities

p, p_i, p_{it}	probability that $\mathcal{I}$ wins a point on service
p^*	optimal/efficient value of p
$\bar{p}$	average p over all matches in a tournament
g, g_i	probability that $\mathcal{I}$ wins a service game
$\tilde{g}_i$	probability that $\mathcal{I}$ wins a service game under the no-ad rule
s_i	probability that $\mathcal{I}$ wins a set
m, m_i	probability that $\mathcal{I}$ wins the match
$m_i(a, b)$	probability that $\mathcal{I}$ wins the match from set score a-b
m^*	value of m if $\mathcal{I}$ serves optimally
v, v_i	probability that team $\mathcal{I}$ wins a rally on service (in volleyball)
v_{ij}	$(1 - v_i)(1 - v_j)$ (in volleyball)

Score probabilities and importance

a-b	point, game, or set score of $\mathcal{I}$ (score a) against $\mathcal{J}$ (score b)
imp, imp_{pm}, imp_{it}	importance of point in match
imp_{pg}, $imp_{pg}(a,b)$	importance of point score a-b in game
imp_{gs}, $imp_{gs}(a,b)$	importance of game score a-b in set
$imp_{gs}(n)$	importance of game number n in set
imp_{sm}, $imp_{sm}(a,b)$	importance of set score a-b in match
$\ell(a,b)$	probability of reaching game score a-b in set
$\ell(n)$	probability that game number n occurs in set

Point scores in a game are counted as 0, 15, 30, 40, advantage, while games and sets are counted as 0, 1, 2, ...

Service probabilities

w_1	probability of winning a point on first service
w_2	probability of winning a point on second service
$w(x)$	probability of winning a point on service using service of type x
x	probability that the service is in
x_1, x_{1i}, x_2, x_{2i}	probability that the first (second) service is in
y_1, y_{1i}, y_2, y_{2i}	probability of winning a point on first (second) service given that the service is in
$y(x)$	probability of winning a point on service given that the service of type x is in
x_1^*, x_2^*, y_1^*, y_2^*	optimal/efficient values corresponding to x_1, x_2, y_1, y_2, respectively

Quality

$bonus_i$	quality gain of low-ranked player $\mathcal{I}$ who has progressed further in the tournament than expected
$malus_i$	quality loss of high-ranked player $\mathcal{I}$ in the early stage of the tournament
q_i	(observed) quality of player $\mathcal{I}$
r_i	stage in the tournament in which player $\mathcal{I}$ is expected to lose, based only on $rank_i$

$rank_i$	official (ATP/WTA) ranking of player $\mathcal{I}$
$round_i$	round in which $\mathcal{I}$ plays current match

Operators

E	expectation
var	variance
cov	covariance
sd	standard deviation
se	standard error
$\sum$	summation operator

Miscellaneous variables

b_i	decimal betting odds for $\mathcal{I}$
d_{it}	dynamic determinant of p_{it}
$diff$	difference between two probabilities
f, f_i, f_{it}	relative frequency that $\mathcal{I}$ wins a point on service
M	number of moment restrictions in GMM estimation
N	total number of matches in the sample
n	game number in a set
T_i	number of points served by $\mathcal{I}$ in the match
T	total number of points in the sample

Random/unexplained parts

π, π_i	unexplained quality part of p, p_i
ϕ, ϕ_i, ϕ_{it}	random noise in f, f_i, f_{it}, respectively

Note that we use Greek letters for random variables corresponding to the equivalent Latin-letter variables.

Parameters

α_b, α_m	impact of $bonus_i$ and $malus_i$ on q_i
β_0	mean of p_i
β_-, β_+	impact of $r_i - r_j$ and $r_i + r_j$ on p_i
β_S, β_R	impact of r_i (server) and r_j (receiver) on ace and double fault probabilities

$\gamma_0, \gamma_1, \lambda$	intercept, slope, and curvature of the $y(x)$-curve
δ_0	mean impact of a dynamic variable on p_{it}
δ_-, δ_+	effect of $r_i - r_j$ and $r_i + r_j$ on the impact of a dynamic variable d_{it} on p_{it}
ρ	correlation coefficient between π_i and π_j (correlation between opponents)
σ^2	variance of π_i (player heterogeneity)

Miscellaneous symbols

$\circ$	not significant
$*$	in a table: significant at 5% level; attached to a probability: optimal/efficient value
$\hat{}$	estimator, estimate
m	subscript when match is the unit of analysis

C

Data, software, and mathematical derivations

In this appendix we draw the reader's attention to our respective websites

www.uva.nl/profile/f.j.g.m.klaassen

and

www.janmagnus.nl/misc/wimbledon.pdf,

where we provide background material. This material includes the complete data set underlying most of our analyses; the computer program, called *Richard*, which we use to calculate probabilities and importances; and also derivations of some of the mathematical results.

Data

The principal (but not the only) data set used in this book consists of point-by-point data of men's and women's singles matches at Wimbledon 1992–1995. These data were given to us by IBM UK, and they were supplemented by ranking data, which we received from the ITF. The data set is described and applied in Chapter 5. Chapters 6–9 use match data, aggregated from these point data. The point data themselves are used again in Chapters 10–12.

The complete data set is presented as an Excel file. This file also contains point-by-point data of the three matches (Federer-Nadal, Clijsters-Williams, Djokovic-Nadal) investigated in Chapters 3 and 4. Included in the Excel file is a separate sheet where the user can easily calculate most of the simple estimates presented in the book, that is, the estimates that do not rely on the GMM estimation method. GMM estimation requires external statistical software.

As an example, consider the relative frequency of winning a point on service for the men. The overall estimate of 64.4% and the standard error of 0.2%, presented on page 71, is replicated in the file. The output automatically provides the breakdown of this frequency, based on whether the server and receiver are seeded or non-seeded players, as reported in Table 7.1. If the user wishes to analyze only a subsample, for example excluding points in a tiebreak and after a break in the previous game within the same set (as used in Table 12.3), he or she can enter this request as well. Precise instructions and explanations are provided in the file.

At several places in the book we also use other data. Summary statistics of the grand slam tournaments from 1992–2010 were provided by IBM UK and the ITF. Data from OnCourt are used in analyzing upsets and the Isner-Mahut match in Chapter 2, and in performing sensitivity analyses for two of the three matches in Chapter 3. The bookmakers' odds at the beginning of each of the three matches were downloaded from Tennis-data.

More tennis data can be obtained from various sources. The websites of the grand slam tournaments provide data and illustrations using IBM SlamTracker. The ATP and WTA websites contain data on many tournaments and rankings. Other websites publish data as well, and their number is growing.

Software: program *Richard*

Richard is a computer program. It computes the probabilities of winning the game (tiebreak), set, and match at any point of a tennis match. The program also produces functions of these probabilities, such as the importance of a game in the set or the importance of a point in the match.

The required inputs are the service point winning probabilities p_i and p_j of the two players $\mathcal{I}$ and $\mathcal{J}$, the current score, the current server, and the rules of the tournament (say, best-of-five sets with a tiebreak in all sets except the final set), but nothing more. The program is fast, accurate, has a convenient 'matrix' setup, is flexible for studying rule changes, and it is freely available.

Richard was born in 1994. It was first written in a computer language called Turbo Pascal. Since Turbo Pascal is not much used any more, we present the program in Matlab, which is more common and more user-friendly.

The program, including all source code, can be downloaded from our websites. Also available on the websites is the associated 'inversion' program, described on page 34, which is important for in-play forecasting and betting. The inversion program produces robust values of p_i and p_j, based on two inputs: the match-winning probability at the beginning of the match (obtained from an outside source) and the sum $p_i + p_j$.

We emphasize three features of *Richard*. First, it is based on the assumption that service points are independent and identically distributed (iid), so that two probabilities (p_i and p_j) suffice. Since tennis has a hierarchical structure (points form games and tiebreaks, which form sets, which form matches), games, tiebreaks and sets are also iid. This allows for a step-by-step calculation of the probability of winning a match. First calculate the game (tiebreak) winning probability, then the set probability, and finally the match probability.

A second distinguishing feature of *Richard* concerns the calculation of the winning probabilities at each step. For example, consider a game served by player $\mathcal{I}$. At score 0-0, $\mathcal{I}$ can win or lose the point, giving 15-0 or 0-15, with probabilities p_i and $1 - p_i$, respectively. After 15-0 the score can be 30-0 or 15-15, with the same transition probabilities, and similarly after 0-15. Special scores are deuce and advantage. Deuce is equivalent to 30-30, and advantage server (receiver) is equivalent to 40-30 (30-40). The game continues until either $\mathcal{I}$ or $\mathcal{J}$ has won the game. We thus obtain seventeen different scores and at each score the probability of the next score is constant: either p_i or $1-p_i$. This type of structure is known as a 'finite Markov chain' (Kemeny and Snell, 1960, pp. 161–167), which has many important and useful properties. One of these properties is that the whole process can be represented in one square of numbers, called a 'matrix'. This matrix has seventeen rows and seventeen columns, and each cell contains the probability of going from one score to the other. Given this matrix, it is then easy to calculate the probability of winning the game from each score. The matrix approach makes the program fast: a few hundredths of a second to compute one match-winning probability. The total computing time for all 413 points in the Federer-Nadal Wimbledon 2008 final is only two seconds on a standard desktop computer.

Third, the program is not only fast, but also flexible. Winning probabilities under non-standard rules (match tiebreak, four-game set, no-ad rule), as discussed on page 27, can be obtained by just changing one single number. Because the full source code is available, the user can adjust the code, if needed, to investigate other, more exotic, rule changes.

Mathematical derivations

For almost all mathematical results in the book we have explained how they were derived and why we needed them. For the remaining formulas we provide full derivations on our websites. These remaining formulas are the following.

Chapter 2

(a) In Section 'From point to game' we consider a game where player $\mathcal{I}$ is serving with constant probability p_i of winning a point. The probability that he or she wins the game is expressed by the formula for g_i on page 15.

(b) Section 'Long matches: Isner-Mahut 2010' (pages 24–27) contains a three-step derivation. This derivation is complete, but the reader may wish to see the explicit expressions (and their derivations) of $\ell(5,5)$ and $\ell(a,b)$. These are provided on the websites.

(c) In Section 'Rule changes: the no-ad rule' we consider the situation where the traditional scoring system at deuce is replaced by the no-ad rule, so that only one deciding point is played at deuce, and we state that the probability that $\mathcal{I}$ wins the game would change from g_i to $\tilde{g}_i$ on page 28.

Chapter 4

In Section 'Big points in a game' (pages 50–52) we consider again a game where player $\mathcal{I}$ is serving with probability p_i of winning a service point. Then we present formulas for $imp_{pg}(40, 30)$, the importance of the score 40-30, and for $imp_{pg}(30, 40)$, the importance of the score 30-40.

Chapter 9

At the end of Section 'Impact on the paycheck' (page 153) we state that if the paycheck doubles in each round, then the expected paycheck for the efficient player will rise by 18.7% for men and by 32.8% for women.

Bibliography

There exists an extensive literature on tennis and tennis statistics. This bibliography only provides the references to the literature discussed in the 'further reading' sections. It does not attempt to cover the whole literature.

Abramitzky, R., L. Einav, S. Kolkowitz, and R. Mill (2012). On the optimality of line call challenges in professional tennis, *International Economic Review*, 53, 939–963.

Albert, J.H. (1993). Comment on 'A statistical analysis of hitting streaks in baseball' by S.C. Albright, *Journal of the American Statistical Association*, 88, 1184–1188.

Albert, J., J. Bennett, and J.J. Cochran (Eds) (2005). *Anthology of Statistics in Sports*, ASA-SIAM Series on Statistics and Applied Probability, SIAM, Philadelphia, PA and ASA, Alexandria, VA.

Albright, S.C. (1993). A statistical analysis of hitting streaks in baseball (including comments and rejoinder), *Journal of the American Statistical Association*, 88, 1175–1196.

Anderson, L.R. (1982). Sex differences on a conjunctive task: Mixed-doubles tennis teams, *Personality and Social Psychology Bulletin*, 8, 330–335.

Bar-Eli, M., S. Avugos, and M. Raab (2006). Twenty years of 'hot hand' research: Review and critique, *Psychology of Sport and Exercise*, 7, 525–553.

Barnett, T. (2006). *Mathematical Modelling in Hierarchical Games with Specific Reference to Tennis*, PhD thesis, Swinburne University of Technology, Melbourne.

Bedford, A., T. Barnett, G.H. Pollard, and G.N. Pollard (2010). How the interpretation of match statistics affects player performance, *Journal of Medicine and Science in Tennis*, 15, 23–27.

Blackman, S.S. and J.W. Casey (1980). Development of a rating system for all tennis players, *Operations Research*, 28, 489–502.

Boulier, B.L. and H.O. Stekler (1999). Are sports seedings good predictors?: An evaluation, *International Journal of Forecasting*, 15, 83–91.

Brody, H., R. Cross, and C. Lindsey (2002). *The Physics and Technology of Tennis*, The United States Racquet Stringers Association, Vista, CA.

Clarke, S.R. (2011). Rating non-elite tennis players using team doubles competition results, *Journal of the Operational Research Society*, 62, 1385–1390.

Clarke, S.R. and D. Dyte (2000). Using official ratings to simulate major tennis tournaments, *International Transactions in Operational Research*, 7, 585–594.

Clarke, S.R. and J.M. Norman (2012). Optimal challenges in tennis. *Journal of the Operational Research Society*, 63, 1765–1772.

Coate, D. and D. Robbins (2001). The tournament careers of top-ranked men and women tennis professionals: Are the gentlemen more committed than the ladies?, *Journal of Labor Research*, 22, 185–193.

Coe, A. (2000). The balance between technology and tradition in tennis, in: *Tennis Science & Technology* (Eds S.J. Haake and A. Coe), Blackwell Science, Oxford, pp. 3–40.

Collins, B. (2010). *The Bud Collins History of Tennis*, 2nd Edition, New Chapter Press, New York.

Crespo, M., M. Reid, and A. Quinn (2006). *Tennis Psychology: 200+ Practical Drills and the Latest Research*, International Tennis Federation, London.

Croucher, J.S. (1981). An analysis of the first 100 years of Wimbledon tennis finals, *Teaching Statistics*, 3, 72–75.

Croucher, J.S. (1998). Developing strategies in tennis, in: *Statistics in Sport* (Ed J. Bennett), Arnold, London, pp. 157–171.

Davies, M., L. Pitt, D. Shapiro, and R. Watson (2005). Betfair.com: Five technology forces revolutionize worldwide wagering, *European Management Journal*, 23, 533–541.

Del Corral, J. (2009). Competitive balance and match uncertainty in grand-slam tennis: Effects of seeding system, gender, and court surface, *Journal of Sports Economics*, 10, 563–581.

Del Corral, J. and J. Prieto-Rodríguez (2010). Are differences in ranks good predictors for grand slam tennis matches?, *International Journal of Forecasting*, 26, 551–563.

Dobson, S. and J. Goddard (2011). *The Economics of Football*, 2nd Edition, Cambridge University Press, New York.

Dudink, A. (1994). Birth date and sporting success, *Nature*, 368, 592.

Easton, S. and K. Uylangco (2010). Forecasting outcomes in tennis matches using within-match betting markets, *International Journal of Forecasting*, 26, 564–575.

Edgar, S. and P. O'Donoghue (2005). Season of birth distribution of elite tennis players, *Journal of Sports Sciences*, 23, 1013–1020.

Elliott, B., Reid, M., and M. Crespo (2009). *Technique Development in Tennis Stroke Production*, International Tennis Federation, London.

Friedman, M. (1953). *Essays in Positive Economics*, Chicago University Press, Chicago.

Gale, D. (1971). Optimal strategy for serving in tennis. *Mathematics Magazine*, 44, 197–199.

Garicano, L., Palacios-Huerta, I., and C. Prendergast (2005). Favoritism under social pressure, *The Review of Economics and Statistics*, 87, 208–216.

George, S.L. (1973). Optimal strategy in tennis: A simple probabilistic model, *Applied Statistics*, 22, 97–104.

Gilovich, T., R. Vallone, and A. Tversky (1985). The hot hand in basketball: On the misperception of random sequences, *Cognitive Psychology*, 17, 295–314.

González-Díaz, J., O. Grossner, and B.W. Rogers (2012). Performing best when it matters most: Evidence from professional tennis, *Journal of Economic Behavior & Organization*, 84, 767–781.

Goodwill, S.R., S.B. Chin, and S.J. Haake (2004). Aerodynamics of spinning and non-spinning tennis balls, *Journal of Wind Engineering and Industrial Aerodynamics*, 92, 935–958.

Guillaume, M., S. Len, M. Tafflet, L. Quinquis, B. Montalvan, K. Schaal, H. Nassif, F.D. Desgorces, and J.-F. Toussaint (2011). Success and decline: Top 10 tennis players follow a biphasic course, *Medicine & Science in Sports & Exercise*, 43, 2148–2154.

Hall, A.R. (2005). *Generalized Method of Moments*, Oxford University Press, New York.

Hart, A. (1997). *Agatha Christie's Miss Marple: The Life and Times of Miss Jane Marple*, Harper Collins Publishers, London.

Holder, R.L. and A.M. Nevill (1997). Modelling performance at international tennis and golf tournaments: Is there a home advantage?, *The Statistician*, 46, 551–559.

Holtzen, D.W. (2000). Handedness and professional tennis, *International Journal of Neuroscience*, 105, 101–119.

Hsi, B.P. and D.M. Burych (1971). Games of two players, *Applied Statistics*, 20, 86–92.

Humphreys, M. (2011). *Wizardry: Baseball's All-Time Greatest Fielders Revealed*, Oxford University Press, New York.

Jackson, D.A. (1994). Index betting on sports, *The Statistician*, 43, 309–315.

Jackson D. and K. Mosurski (1997). Heavy defeats in tennis: Psychological momentum of random effect, *Chance*, 10, 27–34.

Kahneman, D. (2011). *Thinking, Fast and Slow*, Penguin Books Ltd, London.

Kemeny, J.G. and J.L. Snell (1960). *Finite Markov Chains*, Van Nostrand, Princeton, NJ.

Klaassen, F.J.G.M. and J.R. Magnus (2000). How to reduce the service dominance in tennis? Empirical results from four years at Wimbledon, in: *Tennis Science & Technology* (Eds S.J.

Haake and A.O. Coe), Blackwell Science, Oxford, pp. 277–284.

Klaassen, F.J.G.M. and J.R. Magnus (2001). Are points in tennis independent and identically distributed? Evidence from a dynamic binary panel data model, *Journal of the American Statistical Association*, 96, 500–509.

Klaassen, F.J.G.M. and J.R. Magnus (2003a). Forecasting the winner of a tennis match, *European Journal of Operational Research*, 148, 257–267.

Klaassen, F.J.G.M. and J.R. Magnus (2003b). On the probability of winning a tennis match, *Medicine and Science in Tennis*, 8, No. 3, 10–11.

Klaassen, F.J.G.M. and J.R. Magnus (2003c). Forecasting in tennis, in: *Tennis Science & Technology 2* (Ed S. Miller), International Tennis Federation, London, pp. 333–340.

Klaassen, F.J.G.M. and J.R. Magnus (2008). De kans om een tenniswedstrijd te winnen: Federer-Nadal in de finale van Wimbledon 2007, *Stator*, 9, No. 2, 8–11.

Klaassen, F.J.G.M. and J.R. Magnus (2009). The efficiency of top agents: An analysis through service strategy in tennis, *Journal of Econometrics*, 148, 72–85.

Koning, R.H. (2009). Sport and measurement of competition, *De Economist*, 157, 229–249.

Koning, R.H. (2011). Home advantage in professional tennis, *Journal of Sports Sciences*, 29, 19–27.

Körding, K. (2007). Decision theory: What 'should' the nervous system do?, *Science*, 318, 606–610.

Lake, R.J. (2011). Social class, etiquette and behavioural restraint in British lawn tennis, 1870–1939, *The International Journal of the History of Sport*, 28, 876–894.

Lallemand, T., R. Plasman, and F. Rycx (2008). Women and competition in elimination tournaments: Evidence from professional tennis data, *Journal of Sports Economics*, 9, 3–19.

Larkey, P., R. Smith, and J. Kadane (1989). It's okay to believe in the 'hot hand', *Chance*, 2, 22–30.

Lindsey, G.R. (1961). The progress of the score during a baseball game, *Journal of the American Statistical Association*, 56, 703–728.

Little, A. (1995). *Wimbledon Compendium 1995*, The All England Lawn Tennis and Croquet Club, London.

Magnus, J.R. and F.J.G.M. Klaassen (1995). De professor weet wat commentatoren niet weten, *De Volkskrant*, October 14.

Magnus, J.R. and F.J.G.M. Klaassen (1996). Testing some common tennis hypotheses: Four years at Wimbledon, Center for Economic Research, Discussion Paper 1996-73, Tilburg University, The Netherlands.

Magnus, J.R. and F.J.G.M. Klaassen (1997). Wat tenniscommentatoren niet weten: Een analyse van vier jaar Wimbledon, *Kwantitatieve Methoden*, 18, No. 54, 55–62. Reply in: *Kwantitatieve Methoden*, 18, No. 54, 67.

Magnus, J.R. and F.J.G.M. Klaassen (1999a). The effect of new balls in tennis: Four years at Wimbledon, *The Statistician (Journal of the Royal Statistical Society, Series D)*, 48, 239–246.

Magnus, J.R. and F.J.G.M. Klaassen (1999b). On the advantage of serving first in a tennis set: Four years at Wimbledon, *The Statistician (Journal of the Royal Statistical Society, Series D)*, 48, 247–256.

Magnus, J.R. and F.J.G.M. Klaassen (1999c). The final set in a tennis match: Four years at Wimbledon, *Journal of Applied Statistics*, 26, 461–468.

Magnus, J.R. and F.J.G.M. Klaassen (2008). Myths in tennis, in: *Statistical Thinking in Sports* (Eds J. Albert and R.H. Koning), Chapman & Hall/CRC Press, Boca Raton, FL, Chapter 13, pp. 217–240.

McHale, I. and A. Morton (2011). A Bradley-Terry type model for forecasting tennis match results, *International Journal of Forecasting*, 27, 619–630.

Mehta, R., F. Alam, and A. Subic (2008). Review of tennis ball aerodynamics. *Sports Technology*, 1, 7–16.

Miles, R.E. (1984). Symmetric sequential analysis: The efficiencies of sports scoring systems (with particular reference to those of tennis), *Journal of the Royal Statistical Society, Series B*, 46, 93–108.

Miller, S. (2006). Modern tennis rackets, balls, and surfaces, *British Journal of Sports Medicine*, 40, 401–405.

Morris, C. (1977). The most important points in tennis, in: *Optimal Strategies in Sports* (Eds S.P. Ladany and R.E. Machol), North-Holland Publishing Company, Amsterdam, pp. 131–140.

Newton, P.K. and K. Aslam (2006). Monte Carlo tennis, *SIAM Review*, 48, 722–742.

Newton, P.K. and J.B. Keller (2005). Probability of winning at tennis I: Theory and data, *Studies in Applied Mathematics*, 114, 241–269.

Norton, P. and S.R. Clarke (2002). Serving up some grand slam tennis statistics, *Proceedings of the Sixth Australian Conference on Mathematics and Computers in Sport* (Eds G.L. Cohen and T.N. Langtry), Bond University, Queensland, Australia, 1–3 July 2002, pp. 202–209.

O'Donoghue, P. and B. Ingram (2001). A notational analysis of elite tennis strategy, *Journal of Sports Sciences*, 19, 107–115.

O'Malley, A.J. (2008). Probability formulas and statistical analysis in tennis, *Journal of Quantitative Analysis in Sports*, 4, 1–21.

Paserman, M.D. (2010). Gender differences in performance in competitive environments? Evidence from professional tennis players, Boston University and Hebrew University, mimeo.

Pluim, B., S. Miller, D. Dines, P. Renström, G. Windler, B. Norris, K. Stroia, A. Donaldson, and K. Martin (2007). Sport science and medicine in tennis, *British Journal of Sports Medicine*, 41, 703–704.

Pluim, B. and M. Safran (2004). *From Breakpoint To Advantage: A Practical Guide to Optimal Tennis Health and Performance*, The United States Racquet Stringers Association, Vista, CA.

Pollard, G.H. (1983). An analysis of classical and tie-breaker tennis, *Australian Journal of Statistics*, 25, 496–505.

Pollard, G.H. and K. Noble (2004). Benefits of a new game scoring system in tennis: The 50-40 game, in: *Proceedings of the Seventh Australasian Conference on Mathematics and Computers in Sport* (Eds R. Morton and S. Ganesalingam), Massey University, New Zealand, pp. 262–265.

Pollard, G.N. and D. Meyer (2010). An overview of operations research in tennis, *Wiley Encyclopedia of Operations Research and Management Science* (Ed J.J. Cochran), John Wiley & Sons, New York.

Pollard, G.N., G.H. Pollard, T. Barnett, and J. Zeleznikow (2009). Applying tennis match statistics to increase serving performance during a match in progress, *Journal of Medicine and Science in Tennis*, 14, 16–19.

Pollard, G.N., G.H. Pollard, T. Barnett, and J. Zeleznikow (2010). Applying strategies to the tennis challenge system, *Journal of Medicine and Science in Tennis*, 15, 12–16.

Radicchi, F. (2011). Who is the best player ever? A complex network analysis of the history of professional tennis, *PloS ONE*, 6(2): e17249.

Reifman, A. (2012). *Hot Hand: The Statistics Behind Sports' Greatest Streaks*, Potomac Books, Dulles, VA.

Riddle, L.H. (1988). Probability models for tennis scoring systems, *Applied Statistics*, 37, 63–75. Corrigendum, 490.

Riddle, L.H. (1989). Author's reply to D.A. Jackson, *Applied Statistics*, 38, 378–379.

Rohm, A.J., S. Chatterjee, and M. Habibullah (2004). Strategic measure of competitiveness for ranked data, *Managerial and Decision Economics*, 25, 103–108.

Shmanske, S. and L.H. Kahane (Eds) (2012). *The Oxford Handbook of Sports Economics*, two volumes, Oxford University Press, New York.

Sinnett, S. and A. Kingstone (2010). A preliminary investigation regarding the effect of tennis grunting: Does white noise during a tennis shot have a negative impact on shot perception?, *PLoS ONE*, 5(10): e13148.

Siwoff, S., S. Hirdt, and P. Hirdt (1987). *The 1987 Elias Baseball Analyst*, Collier Books, New York.

Spurr, J. and J. Capel-Davies (2007). Tennis ball durability: Simulation of real play in the laboratory, in: *Tennis Science & Technology 3* (Eds S. Miller and J. Capel-Davies), International Tennis Federation, London, pp. 41–49.

Stefani, R.T. (1997). Survey of the major world sports rating systems, *Journal of Applied Statistics*, 24, 635–646.

Stern, H.S. and C.N. Morris (1993). Comment on 'A statistical analysis of hitting streaks in baseball' by S.C. Albright, *Journal of the American Statistical Association*, 88, 1189–1194.

Sunde, U. (2009). Heterogeneity and performance in tournaments: A test for incentive effects using professional tennis data, *Applied Economics*, 41, 3199–3208.

Tversky, A. and D. Kahneman (1986). Rational choice and the framing of decisions, *Journal of Business*, 59, 251–278.

Walker, M. and J. Wooders (2001). Minimax play at Wimbledon, *American Economic Review*, 91, 1521–1538.

Walker, M., J. Wooders, and R. Amir (2011). Equilibrium play in matches: Binary Markov games, *Games and Economic Behavior*, 71, 487–502.

Wozniak, D. (2012). Gender differences in a market with relative performance feedback: Professional tennis players, *Journal of Economic Behavior & Organization*, 83, 158–171.

Index

11704334R00160

Printed in Great Britain
by Amazon.co.uk, Ltd.,
Marston Gate.